AF779873

Development of Boron-Based Materials

Development of Boron-Based Materials

Guest Editor

Nevill Gonzalez Szwacki

MDPI

Basel • Beijing • Wuhan • Barcelona • Belgrade • Novi Sad • Cluj • Manchester

Guest Editor
Nevill Gonzalez Szwacki
Faculty of Physics
University of Warsaw
Warsaw
Poland

Editorial Office
MDPI AG
Grosspeteranlage 5
4052 Basel, Switzerland

This is a reprint of the Special Issue, published open access by the journal *Materials* (ISSN 1996-1944), freely accessible at: https://www.mdpi.com/journal/materials/special_issues/5N2E80UNHF.

For citation purposes, cite each article independently as indicated on the article page online and as indicated below:

Lastname, A.A.; Lastname, B.B. Article Title. *Journal Name* **Year**, *Volume Number*, Page Range.

ISBN 978-3-7258-6135-4 (Hbk)
ISBN 978-3-7258-6136-1 (PDF)
https://doi.org/10.3390/books978-3-7258-6136-1

© 2026 by the authors. Articles in this book are Open Access and distributed under the Creative Commons Attribution (CC BY) license. The book as a whole is distributed by MDPI under the terms and conditions of the Creative Commons Attribution-NonCommercial-NoDerivs (CC BY-NC-ND) license (https://creativecommons.org/licenses/by-nc-nd/4.0/).

Contents

About the Editor

Nevill Gonzalez Szwacki

Nevill Gonzalez Szwacki is an Assistant Professor at the Faculty of Physics, University of Warsaw. His research focuses on first-principles modeling of low-dimensional and boron-related materials, with particular emphasis on their electronic, magnetic, and structural properties. He is internationally recognized for the prediction of boron fullerenes, including the B_{80} "boron buckyball," and has contributed significantly to the understanding of boron allotropes in various dimensions. He has published over 50 peer-reviewed papers, edited two Special Issues (*Materials* and *Nanomaterials*), and served as a reviewer for numerous international journals. His academic career has included roles at the University of Warsaw, Rice University, Texas Tech University, and the Polish Academy of Sciences.

Preface

This Reprint, "Development of Boron-Based Materials", reflects the rapid growth and diversification that has occurred in boron research. The present collection aims to provide a unified view of how boron's structural versatility translates into practical applications in materials science and technology. By gathering high-quality contributions from international experts, the Reprint addresses both the fundamental understanding and the technological potential of boron-rich systems. Topics range from nanostructured borophenes and borometallic clusters to boron-based composites, glasses, and carbides. The Reprint is addressed to researchers and graduate students in chemistry, physics, and engineering who seek a broad and up-to-date overview of boron materials science. We hope that it will inspire further advances in this fascinating field.

Nevill Gonzalez Szwacki
Guest Editor

materials

Editorial

Development of Boron-Based Materials

Nevill Gonzalez Szwacki

Faculty of Physics, University of Warsaw, Pasteura 5, 02093 Warsaw, Poland; gonz@fuw.edu.pl;
Tel.: +48-22-55-32-797

The field of boron-based materials has undergone a significant transformation in recent decades, propelled by the unique chemistry of boron and its ability to form complex, multi-dimensional structures with exceptional physical, chemical, and electronic properties. With its electron deficiency and strong covalent bonding tendencies, boron forms a vast range of clusters, layers, and networks that do not find parallels among most other elements [1]. This Special Issue, entitled "Development of Boron-Based Materials", brings together nine original research and review articles that collectively reflect the rapid expansion and diversification of the science of boron materials. These contributions represent a cross-section of the state-of-the-art approaches to synthesizing, characterizing, and theoretically modeling boron-rich compounds and composites with potential applications in electronics, energy storage, radiation shielding, catalysis, and biomedical science.

There are a range of motivations for investigating boron-based materials. For one, the light weight and abundance of boron make it attractive for sustainable applications. Furthermore, its structural versatility and ability to host metallic, semiconducting, insulating, and superconducting phases ensure a broad platform for technological innovations. From bulk ceramics and metallic borides to low-dimensional nanostructures and glasses, boron-containing systems demonstrate high hardness, chemical inertness, tunable band gaps, and exceptional thermal stability. These features, in turn, enable potential applications in armor systems, solid-state batteries, thermoelectrics, high-temperature electronics, and even quantum materials. Furthermore, current advances in computational modeling allow us to predict and screen novel boron configurations with a degree of precision that complements experimental synthesis and characterization.

Several contributions in this Special Issue focus on low-dimensional and nanoscale boron-based systems, which are at the forefront of condensed matter physics and material engineering. Arias-Camacho and Gonzalez Szwacki [2] present a computational study on the structural, electronic, magnetic, and transport properties of two-dimensional transition metal monoborides—specifically Cr, Fe, and Zr-based MBenes. These monolayer materials are theoretically stable in orthorhombic and hexagonal symmetries and display metallic behavior. Interestingly, significant magnetic moments in some of these configurations point to their potential utility as 2D magnets. Further extending the frontiers of two-dimensional boron science, Zhong et al. [3] investigate bilayer borophene, demonstrating a unique coexistence of superhard mechanical properties and tunable superconducting behavior. Their theoretical analysis shows that bilayer-δ_6 borophene has a very high Young's modulus and can reach a superconducting transition temperature up to 46 K under strain. These characteristics place it among the most promising 2D materials for mechanical and quantum applications. Yin et al. [4] contribute an experimental study on hydrogen boride (HB) sheets, focusing on the adsorption of atomic hydrogen. Using photoelectron spectroscopy, the authors identify chemical transformations at the boron surface, indicative of potential use in hydrogen storage and sensor technologies. These HB sheets are especially notable for their aqueous stability and functional surface chemistry. Perveen and Gonzalez Szwacki [5]

 https://doi.org/10.3390/ma18102247

explore the properties of borometallic molecular wheel clusters, where transition metal atoms are embedded in planar or drum-like boron ring structures. These clusters, predicted via first-principles calculations, are shown to be electronically stable and possess nontrivial magnetic properties. Their tunability suggests they may serve as fundamental units in future nanoscale spintronic devices.

Beyond low-dimensional systems, this Special Issue features studies that examine bulk boron-rich compounds for use in harsh environments. Iwan et al. [6] report the synthesis of a boron-rich boron carbide compound, $B_{4.55}C$, via spark plasma sintering. Characterized by high hardness, thermal oxidation resistance, and neutron absorption capability, this material is suitable for high-temperature structural applications in aerospace and nuclear systems. In a more fundamental investigation, Werheit [7] explores the nature of phase transitions in boron carbide. His study identifies temperature- and pressure-induced transitions closely tied to subtle shifts in bonding and structural defects. These transformations are critical for understanding the anomalous electronic and thermal transport behaviors observed in boron-rich carbides. This Special Issue also explores functional boron-containing networks rooted in cluster and glass chemistry. Avdeeva et al. [8] provide a comprehensive review on the use of boron cluster anions and carboranes to generate advanced materials such as boron carbide, boron nitride, and metal borides. Their utility in protective coatings, composites, and neutron shielding materials is emphasized, showing the synthetic versatility of boron-rich precursors.

The protective capabilities of boron-based materials are well represented by the work of Avcıoğlu and Avcıoğlu [9], who evaluate transition metal borides as candidates for all-in-one radiation shielding. Using advanced simulation tools, they analyze gamma ray and neutron attenuation efficiencies. Compounds such as ReB_2 and SmB_6 are found to exceed the shielding performance of traditional lead-based materials, offering a more sustainable and less toxic alternative for shielding technologies in nuclear and aerospace contexts.

Finally, Kettlewell and Boyd [10] address the challenges of predicting structure–property relationships in oxyhalide borate glasses. These glasses, which operate in the complex "borate anomaly" regime, are modeled using a statistical design of mixtures methodology. Their study provides predictive frameworks for tuning degradation, ion release, and thermal stability—key features for biomedical and optoelectronic applications.

Altogether, the nine papers in this Special Issue offer a panoramic view of the current research landscape in boron-based materials. This collection demonstrates the diversity and depth of modern boron science, from the atomistic design of new 2D structures and bulk ceramics to the development of bioactive glasses and multifunctional composites. It highlights the synergies between theory and experiment, between chemistry and engineering, and between fundamental insights and applied objectives.

Acknowledgments: As Guest Editor, I sincerely thank all authors, reviewers, and editorial staff for their contributions to this Special Issue. I am confident that this compilation will serve as a reference point and inspiration for researchers exploring the infinite opportunities within boron-based materials.

Conflicts of Interest: The authors declare no conflicts of interest.

References

1. Matsuda, I.; Wu, K. (Eds.) *2D Boron: Boraphene, Borophene, Boronene*; Springer International Publishing: Cham, Switzerland, 2021. [CrossRef]
2. Arias-Camacho, I.M.; Gonzalez Szwacki, N. Exploring the Structural, Electronic, Magnetic, and Transport Properties of 2D Cr, Fe, and Zr Monoborides. *Materials* **2023**, *16*, 5104. [CrossRef] [PubMed]
3. Zhong, C.; Sun, M.; Altalhi, T.; Yakobson, B.I. Superhard and Superconducting Bilayer Borophene. *Materials* **2024**, *17*, 1967. [CrossRef] [PubMed]

4. Yin, H.; Tang, J.; Yamaguchi, K.; Sakurai, H.; Tsujikawa, Y.; Horio, M.; Kondo, T.; Matsuda, I. Adsorption of Atomic Hydrogen on Hydrogen Boride Sheets Studied by Photoelectron Spectroscopy. *Materials* **2024**, *17*, 4806. [CrossRef] [PubMed]
5. Perveen, S.; Gonzalez Szwacki, N. Structural, Electronic, and Magnetic Properties of Neutral Borometallic Molecular Wheel Clusters. *Materials* **2025**, *18*, 459. [CrossRef] [PubMed]
6. Iwan, S.; Sutton, W.; Baker, P.A.; Sereika, R.; Vohra, Y.K. Synthesis and Thermal Oxidation Resistance of Boron-Rich Boron–Carbide Material. *Materials* **2023**, *16*, 6526. [CrossRef] [PubMed]
7. Werheit, H. Phase Transitions in Boron Carbide. *Materials* **2023**, *16*, 6734. [CrossRef] [PubMed]
8. Avdeeva, V.V.; Nikiforova, S.E.; Malinina, E.A.; Sivaev, I.B.; Kuznetsov, N.T. Composites and Materials Prepared from Boron Cluster Anions and Carboranes. *Materials* **2023**, *16*, 6099. [CrossRef]
9. Avcıoğlu, C.; Avcıoğlu, S. Transition Metal Borides for All-in-One Radiation Shielding. *Materials* **2023**, *16*, 6496. [CrossRef] [PubMed]
10. Kettlewell, B.; Boyd, D. Inside the Borate Anomaly: Leveraging a Predictive Modelling Approach to Navigate Complex Composition–Structure–Property Relationships in Oxyhalide Borate Glasses. *Materials* **2024**, *17*, 2073. [CrossRef]

Disclaimer/Publisher's Note: The statements, opinions and data contained in all publications are solely those of the individual author(s) and contributor(s) and not of MDPI and/or the editor(s). MDPI and/or the editor(s) disclaim responsibility for any injury to people or property resulting from any ideas, methods, instructions or products referred to in the content.

materials

MDPI

Article

Exploring the Structural, Electronic, Magnetic, and Transport Properties of 2D Cr, Fe, and Zr Monoborides

Isabel M. Arias-Camacho and Nevill Gonzalez Szwacki *

Faculty of Physics, University of Warsaw, Pasteura 5, PL-02093 Warsaw, Poland; isabel.arias@fuw.edu.pl
* Correspondence: gonz@fuw.edu.pl; Tel.: +48-22-55-32-797

Abstract: Compared to other 2D materials, MBenes are at an early stage of investigation in terms of both experimental and theoretical approaches. However, their wide range of possible 2D structures leads to novel and challenging properties and consequent applications. From all the possible stoichiometries, we performed a theoretical study of orthorhombic and hexagonal M_2B_2 MBenes within the framework of density functional theory. We found that both symmetries of Cr_2B_2, Fe_2B_2, and Zr_2B_2 show metallic behavior and could be grown under certain conditions as they were demonstrated to be dynamically stable. Moreover, the values of the magnetic moment observed, in specific ferromagnetic cases exceeding 2.5 μ_B/M_2B_2, make them suitable as robust 2D magnets. Our findings represent an important step in the understanding of MBenes and open several windows to future research in fields like energy conversion and storage, sensing, catalysis, biochemistry, and nanotechnology, among others.

Keywords: transition metal borides; MBenes; energy conversion and storage; 2D magnets

1. Introduction

Transition metal borides (MBs) can be regarded as new efficient earth-abundant materials for energy storage/conversion systems, including metal-ion batteries, metal-air batteries, capacitors, oxygen evolution reactions (OERs), and other electrochemical fields [1]. Single-crystalline ternary transition metal borides (MAB phases, where M are transition metals, A are *p*-block elements, and B is boron) were first reported in 2015 by Ade and Hillebrecht and recently gained attention as promising layered materials [2]. Their two-dimensional counterparts (MBenes) can be obtained from the chemical exfoliation of the MAB phases. It is noted that MBenes possess different stoichiometries and variable modes of 2D layer sandwiching compared to the corresponding MXenes [3].

The MAB phases with M and B with one-to-one stoichiometry are orthorhombic and hexagonal crystals with the chemical formulae of MAB and M_2AB_2 [3]. Experimentally, orthorhombic MAlB (M = Mo and W), M_2AlB_2 (M = Cr, Mn, and Fe), and hexagonal Ti_2InB_2 [4] have already been synthesized. The MAB phases are promising candidates for obtaining new 2D MBenes. A 2D MoB was reported to have been obtained by the partial etching of Mo_2AlB_2 phases through the deintercalation of Al layers from the ordered stacking faults region [5,6]. Selectively HCl-etching Al layers from Cr_2AlB_2 yielded bulk-layered CrB nanosheets [7,8]. Removal of the indium layer through the high-temperature dealloying of Ti_2InB_2 yielded a bulk-layered TiB structure [4]. To date, however, the synthesis of individual single-layer MBenes has not been realized.

The present work concerns computational studies of the structural, energetic, electronic, and transport properties of selected MBene compounds with M = Cr, Fe, and Zr. Some of the investigated 2D structures have been proven to be stable in previous studies [9]. The studied MBenes present either an orthorhombic structure (ortho-MBenes) with *Pmma* (no. 51) space group symmetry or a hexagonal structure (hex-MBenes) with *P6/mmm* (no. 191) space group symmetry. In the *Pmma* structures, each atom is surrounded by six neighbors, and the buckled bilayers are sandwiched between transition metal (TM) layers. On the other hand, in the $P6/mmm$ structures, the honeycomb-type boron layer is sandwiched between two TM layers on both sides, and every TM atom is located above or below the centroid of the honeycomb structure.

The bulk counterparts of our metal monoboride nanosheets are the ferromagnetic α and β modifications of FeB and the nonmagnetic CrB and ZrB compounds, all widely studied, both experimentally and theoretically [10–12]. The structure of α-FeB is debatable [12], whereas β-FeB and CrB are orthorhombic crystals with $Pnma$ (no. 62) and $Cmcm$ (no. 63) space group symmetries, respectively. The ZrB solid is rock-salt-structured and crystallizes in a cubic $Fm\overline{3}m$ (no. 225) space group symmetry. The β-FeB and CrB solids exhibit very interesting structures since both enclose boron double-chain (BDC) stripes which are very common motifs of all-boron nanostructures [13,14].

The purpose of this work is to understand the origin of the physical properties of MBene (M = Cr, Fe, and Zr) compounds, as well as to identify features that may affect the transport properties of these compounds.

2. Computational Approach

First-principles spin-polarized calculations were performed within the framework of density functional theory (DFT) within the generalized gradient corrected approximation of Perdew–Burke–Ernzerhof (PBE) [15] for the exchange-correlation functional, using projector plane-wave (PAW) pseudopotentials [16] as implemented in the Quantum ESPRESSO (QE) suite of codes [17]. Every unit cell consists of two atoms of boron and two atoms of the TM. To avoid interactions between adjacent MBenes, we considered an empty space of thickness 15 Å along the normal direction. Optimized geometries were reached allowing the unit cell shape, volume, and the ions to relax until the residual forces on the atoms were less than 0.3 meV/Å and the total energy (E_{tot}) convergence was set to 10^{-5} Ry. We expanded the electronic wave functions and charge density in plane-wave basis sets with an energy cutoff of 70 and 700 Ry, respectively, while the Γ-centered k-point grid in the Brillouin zone, in the Monkhorst–Pack scheme, was set to $12 \times 12 \times 1$ for the geometry optimization and $24 \times 24 \times 1$ for the DOS calculations, with Gaussian smearing of 0.02 Ry; these values ensure the accuracy of E_{tot}. All the structures were considered to be initially spin-polarized and, in order to determine the magnetic ground states of each, we calculated two magnetic configurations—one ferromagnetic (FM) and one antiferromagnetic (AFM)—to reach the most energetically favorable.

For the structural characterization of the studied systems, an important descriptor we use is the cohesive energy per atom (E_{coh}); that is, the difference in energy between E_{tot} of the compound and the sum of the total energies of the isolated atoms,

$$E_{coh} = (E[\mathrm{M_2B_2}] - n_{\mathrm{B}}E[\mathrm{B}] - n_{\mathrm{M}}E[\mathrm{M}])/(n_{\mathrm{B}} + n_{\mathrm{M}}), \quad (1)$$

which means the released energy when a compound dissociates into isolated free atoms, where M represents the TM atom, $E[\mathrm{M_2B_2}]$ is the E_{tot} of the MBene, $E[\mathrm{B}]$ and $E[\mathrm{M}]$ are the total energies of the isolated atoms (B and TM atoms), and n_{B} and n_{M} are the numbers of boron and TM atoms per unit cell, respectively, directly obtained from the spin-polarized calculations. The phonon-dispersion curves were obtained by means of density functional perturbation theory (DFPT), calculating the dynamical matrices in the linear response approach on a q-point grid of $4 \times 4 \times 1$. The transport integrals were computed using Boltzmann transport theory and a constant scattering rate model [18]. The charge transfer was obtained by Bader analysis [19] and all the visualizations were performed using the Visualization for Electronic and STructural Analysis (VESTA) software [20].

3. Results and Discussion

3.1. Structure and Stability

Since boron is electron deficient, it is expected that a mixture with TM atoms will lead to stable structures. As mentioned above, among all the possible MBenes, we focused on those which possess either orthorhombic or hexagonal M_2B_2 structures, which are shown in Figure 1. After a full structural optimization, we found that the unit cells of ortho-MBenes became almost rectangular with $a > b$ when the TM was Fe or Cr (a/b is 1.005 and 1.013 for chromium and iron, respectively), whereas $a < b$ for ortho-Zr_2B_2 ($a/b = 0.94$). All the cell parameters are described in Table 1 and compared with other literature reports.

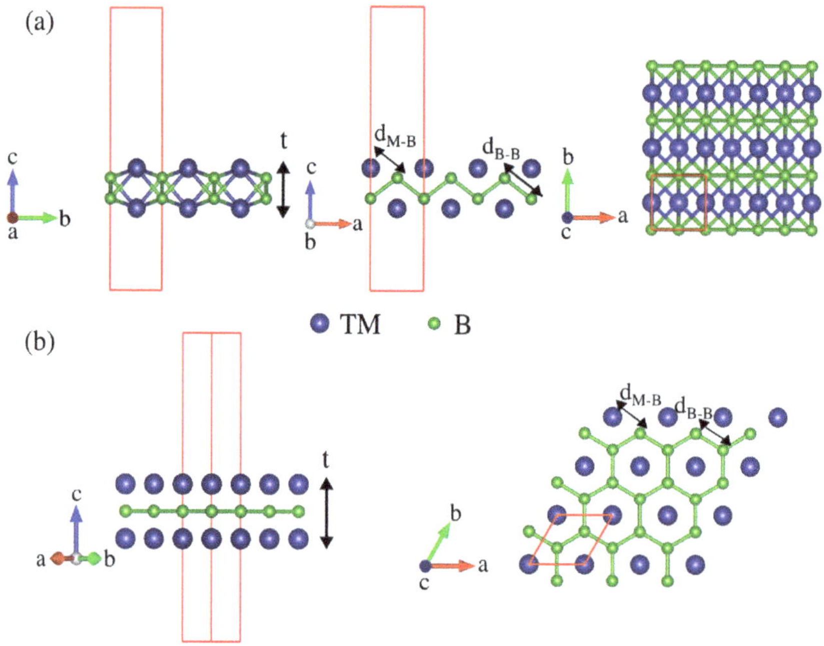

Figure 1. Side (**left**) and top (**right**) views of the (**a**) ortho-MBene and (**b**) hex-MBene structures of M_2B_2 corresponding to *Pmma* and *P6/mmm* symmetries, respectively. The unit cells used in the calculations are shown in red.

The E_{coh} and phonon dispersion curves are good indicators of the bond strength and dynamical stability of the materials. The dynamical matrix gives us the frequency $\omega(\mathbf{q})$ whose square is negative when there exist instabilities for a particular phonon mode with $\mathbf{q}$ (imaginary frequencies), which means that this mode does not generate the restoring force needed by the lattice vibrations and could take the structure away from the original configuration. The results of our calculations are summarized in Table 2 and in Figure 2.

Table 1. Lattice parameters (a and b), boron–boron distance ($d_{B\text{–}B}$), TM–boron distance ($d_{M\text{–}B}$), and thickness of the 2D structure (t) for Cr_2B_2, Fe_2B_2, and Zr_2B_2. The $d_{B\text{–}B}$, $d_{M\text{–}B}$, and t parameters are defined in Figure 1. All the values are in Angstroms (Å) and, for comparison purposes, the data in brackets are taken from the literature.

	Cr_2B_2		Fe_2B_2		Zr_2B_2	
	Pmma	*P6/mmm*	*Pmma*	*P6/mmm*	*Pmma*	*P6/mmm*
a	2.885 (2.860 [4]) (2.930 [5])	2.919 (2.921 [1]) (2.926 [2])	2.823 (2.800 [4]) (2.770 [5])	2.913	3.084 (3.07 [7])	3.159 (3.144 [1]) (3.160 [2]) (3.134 [3])

Table 1. *Cont.*

	Cr_2B_2		Fe_2B_2		Zr_2B_2	
b	2.870 (2.850 [4]) (2.870 [5])	2.919 (2.921 [1]) (2.926 [2])	2.787 (2.680 [4]) (2.820 [5])	2.913	3.281 (3.27 [7])	3.159 (3.144 [1]) (3.160 [2]) (3.134 [3])
$d_{B\text{–}B}$	1.810	1.685 (1.689 [2])	1.813	1.682	1.721 (1.71 [7])	1.824 (1.825 [2]) (1.891 [3])
$d_{M\text{–}B}$	2.10 (2.11 [4])	2.15	2.04 (2.05 [4])	2.15	2.46 (2.43 [7])	2.49 (2.49 [3])
t	2.122	2.662 (2.651 [1]) (2.639 [2])	2.134 (2.13 [6])	2.680	2.817 (2.84 [7])	3.382 (3.391 [1]) (3.380 [2]) (3.385 [3])

[1] Ref. [21]. [2] Ref. [22]. [3] Ref. [23]. [4] Ref. [24]. [5] Ref. [25]. [6] Ref. [26]. [7] Ref. [27].

Table 2. Cohesive energy (E_{coh}), highest frequency at the Γ point (ν), and charge transfer from TM to B (Δq) for Cr_2B_2, Fe_2B_2, and Zr_2B_2. The values in bold are to highlight the structure with higher E_{coh}.

	Cr_2B_2		Fe_2B_2		Zr_2B_2	
	Pmma	*P6/mmm*	*Pmma*	*P6/mmm*	*Pmma*	*P6/mmm*
E_{coh} (eV)	**6.222**	6.201	**6.901**	6.830	8.050	**8.087**
ν (cm^{-1})	762.45	1003.50	772.74	976.01	928.02	744.12
Δq (*e*)	−0.76	−0.61	−0.37	−0.40	−1.17	−0.86

All our MBenes exhibit large E_{coh} values ranging from 6.222 to 8.087 eV, as shown in Table 2, which means they present strong internal binding and good stability. Moreover, our results are in good agreement with previous works. For instance, Zhang et al. [28] obtained a value of 6.30 eV for ortho-Cr_2B_2. For the sake of comparison, we also computed the diamond structure of carbon using the same optimization parameters, resulting in a E_{coh} of 7.757 eV, which is a comparable value to that of other theoretical and experimental reports [29]. Interestingly, other studies have noted a dependence of the structure stability with the atomic mass of the TM [22], a fact that is also reproduced in our results, hex-Zr_2B_2 being the MBene with the highest E_{coh} (8.087 eV). From a structural point of view, Cr_2B_2 and Fe_2B_2 prefer to adopt orthorhombic structures, whereas Zr_2B_2 accommodates better to the hexagonal one. However, the orthorhombic and hexagonal phases are close in energy (within some tens of meV), and, according to recent reports [9,30], ortho-MBenes might transform into hex-MBenes at high temperatures.

On the one hand, calculations of the phonon frequencies reveal that some small imaginary frequencies appear in the surroundings of the Γ point for orthorhombic Cr_2B_2 and Zr_2B_2 as shown on the left panel of Figure 2. On the other hand, none of the hexagonal structures have imaginary frequencies as shown on the right panel of Figure 2. Similar dynamical instabilities as for ortho-MBenes have been reported for freestanding 2D structures in previous works [25,31]. These studies have concluded that the out-of-plane acoustic mode, ZA, is responsible for such instabilities, which are against the long-wavelength transversal waves that could be fixed by defects like grain boundaries or ripples [32].

The highest values of the frequencies at the Γ-point are collected in Table 2. All the frequencies are higher than 740 cm^{-1}. Starting with the orthorhombic structures, the highest optical frequency at Γ-point increases with the atomic number of the TM atom; that is, Cr, Fe, and Zr, in that order. These results are very close to those found in the literature [24]. Since the optical frequency is an indicator of the bond strength, Zr_2B_2 is more stable than Cr_2B_2 and Fe_2B_2. The trend is, however, the opposite for hex-mBenes, meaning that the highest frequency at the Γ-point corresponds to the TM atom which has the smallest atomic number, in this case Cr, and decreases for Fe and Zr. We can conclude that the studied MBenes are stable and could be grown experimentally under certain conditions.

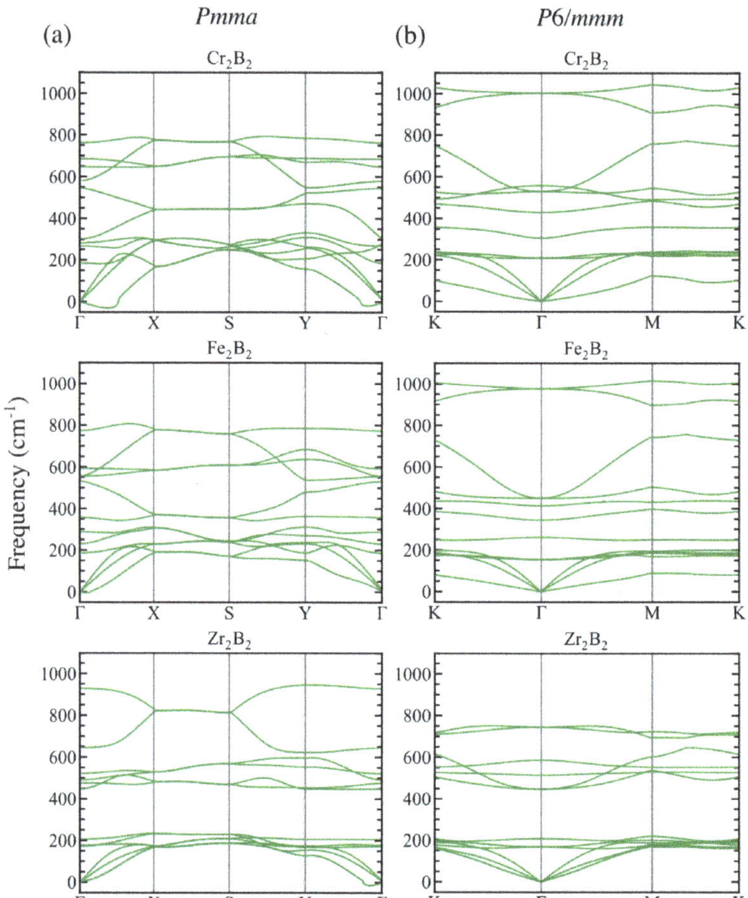

Figure 2. Phonon dispersion relations for the (**a**) *Pmma* and (**b**) *P6/mmm* structures of Cr_2B_2, Fe_2B_2, and Zr_2B_2.

3.2. Electronic Properties

To understand the electronic behavior of the MBenes involved in our study, we calculated the spin-polarized band structure, the density of states (DOS), and the projected density of states (PDOS) for the studied systems. The results of these calculations are shown in Figures 3 and 4 for ortho-MBenes and hex-MBenes, respectively. Looking at the band structure and DOS, we may conclude that there are no band gaps between the valence band (VB) and the conduction band (CB) for any of the studied structures, which means that all the systems are metallic with partially occupied bands crossing the Fermi level (for the majority and the minority spin channels). This metallic character of the pristine MBenes has also been reported in other works [33]. In all cases, the *p* orbitals of boron are deep in energy (ranging from −8 to −2 eV approximately) and, in the orthorhombic structures, partially hybridize with the *d* orbitals of the TM. Near the Fermi level, the PDOS for both symmetries of Cr_2B_2 and Fe_2B_2 is dominated by the *d* orbitals of Cr and Fe, respectively (see the right panels of (a) and (b) in Figures 3 and 4). Whereas, hybridization between the *p* and *d* orbitals of Zr occurs at the Fermi level of Zr_2B_2 (see the right panels of (c) in Figures 3 and 4). The contribution of the majority states at the Fermi level of ortho-Fe_2B_2 is very small, whereas an equal contribution of the minority and majority states is observed in hex-Fe_2B_2 and also in both symmetries of Zr_2B_2. Finally, the contribution of the majority states is larger than that of the minority states at the Fermi level for Cr_2B_2.

According to our Bader analysis, the charge transfer, Δq, always occurs from the TMs to the boron atoms. This is shown in Table 2 where we present the values of Δq for all the studied cases. The obtained values are in agreement with other reports (e.g., $\Delta q = -0.34e$ for ortho-Fe_2B_2, as reported in ref. [26]). In general, considering both the *Pmma* and $P6/mmm$ structures, the largest charge transfer occurs between Zr and B, whereas Fe is the TM for which the charge transfer to B is the smallest one.

The conductivity results are shown in Figure 5. They reveal anisotropy of the conductivity tensor for the ortho-MBenes, especially for the case of ortho-Zr_2B_2 for which the direction perpendicular to the BDC is clearly the preferred one, whereas for ortho-Fe_2B_2, the conductivity is higher along the BDC. On the other hand, the hex-MBenes are isotropic with respect to conductivity. It is also worth highlighting that the orthorhombic Fe_2B_2 and Zr_2B_2 present the highest conductivity values among all the studied cases.

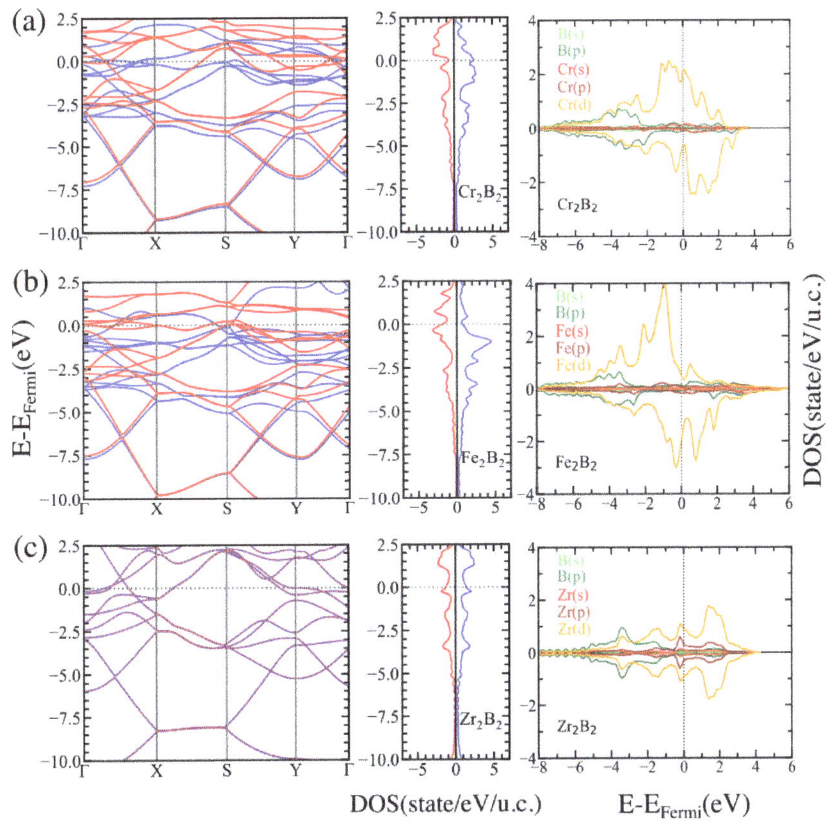

Figure 3. Electronic band structure (**left**) and DOS (**center**) for the majority (blue) and minority (red) spin states, and orbital-resolved PDOS (**right**) for the *Pmma* structures of (**a**) Cr_2B_2, (**b**) Fe_2B_2, and (**c**) Zr_2B_2.

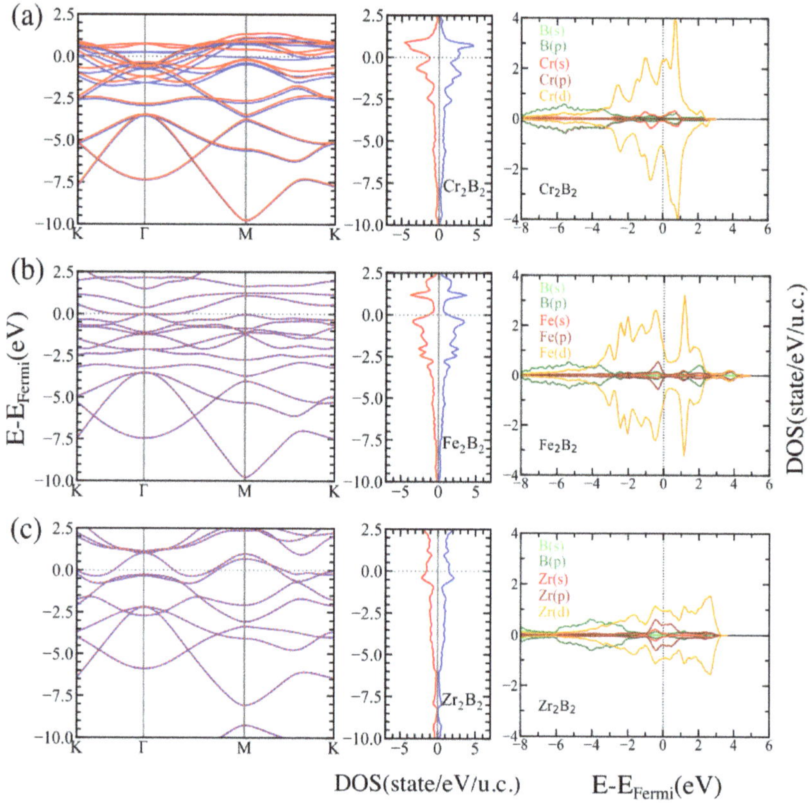

Figure 4. Electronic band structure and DOS for the majority (blue) and minority (red) spins, and orbital-resolved projected density of states (PDOS) for the $P6/mmm$ structures of (a) Cr_2B_2, (b) Fe_2B_2, and (c) Zr_2B_2.

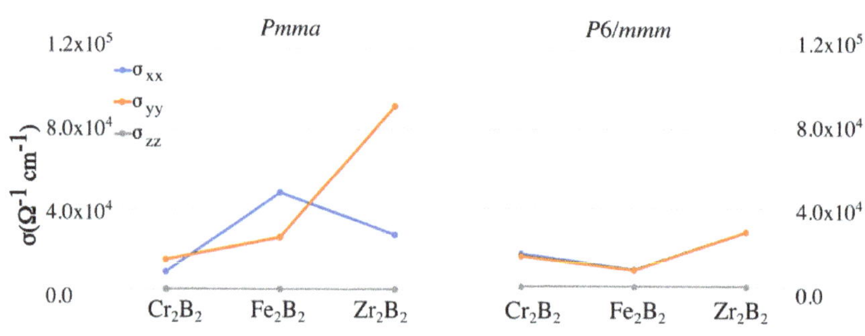

Figure 5. Components of the conductivity tensor for the $Pmma$ (left) and $P6/mmm$ (right) structures of Cr_2B_2, Fe_2B_2, and Zr_2B_2.

3.3. Magnetic Properties

The origin of magnetism in these types of materials arises from the d orbitals of the TM atoms. Both FM and AFM configurations have been suggested to determine the ground state. The results of our calculations are summarized in Table 3. Among all the considered structures, only the hex-Fe_2B_2 resulted in an AFM ground state since the E_{tot} value of the FM state is 46.84 meV higher in energy. On the other hand, ortho-Fe_2B_2 exhibited an FM ground state in accordance with previous reports [34]. However, a more detailed analysis, that also included the next nearest neighbors and was performed using DFT combined with the Monte Carlo method, revealed that ortho-Fe_2B_2 actually has a stable columnar AFM ground state [26]. In the particular case of Cr_2B_2, both structures result in an FM arrangement of the magnetic moments. In general, the total energy difference, $\Delta E_{\text{FM-AFM}}$, between the FM and AFM configurations is always higher in absolute value for the orthorhombic structures (−104.38 and −108.12 eV for Cr_2B_2 and Fe_2B_2, respectively,) than for the hexagonal ones (−0.29 and 46.84 eV for Cr_2B_2 and Fe_2B_2, respectively) indicating that, in the latter case, the magnetic ordering may not be preserved at room temperature. Both orthorhombic Cr_2B_2 and Fe_2B_2 structures show an FM ground state with a magnetic moment over 2.5 μ_B per unit cell, a suitable behavior for robust 2D magnets. Interestingly, the boron atoms are also slightly polarized for those cases for which the TM–boron distances ($d_{\text{M-B}}$) are the shortest (see Table 1). Finally, Zr_2B_2 MBenes exhibit non-magnetic behavior.

The energy difference $\Delta E_{\text{FM-AFM}} = E_{\text{FM}} - E_{\text{AFM}}$ can be used to evaluate the exchange interaction, J_{NN}, between TM atoms at the nearest neighbor (NN) positions. The exchange energy for a system of interacting atomic moments, $\mathbf{S}_i$, can be described by the Heisenberg model:

$$E_{tot} = E_0 - \frac{1}{2}\sum_{i \neq j} J_{ij}\mathbf{S}_i \cdot \mathbf{S}_j, \tag{2}$$

where E_0 is the total energy excluding spin–spin interactions and in our case $S_i = S_j = S$. For ferromagnetically or antiferromagnetically coupled TM ions at NN positions $-2J_{\text{NN}}S^2 = \Delta E_{\text{FM-AFM}}$. The critical temperature (Curie or Néel temperature), T_c, in the mean field approximation (MFA) can be estimated from

$$T_c \simeq \frac{2}{3}k_{\text{B}} \cdot \left(J_{\text{NN}}S^2\right) = \frac{1}{3}k_{\text{B}} \cdot |\Delta E_{\text{FM-AFM}}|. \tag{3}$$

The T_c values are collected in Table 3. As can be seen from the table, we obtain a considerably large value of $T_c = 418$ K for ortho-Fe_2B_2. However, as mentioned above, a more accurate investigation [26] predicts an AFM ground state with $T_c = 115$ K. Interestingly, our calculations predict an AFM ground state for hex-Fe_2B_2 with $T_c = 181$ K.

Table 3. Total magnetic moment (μ_{tot}), magnetic moment of the TM atoms (μ_{TM}), magnetic moment induced on the boron atoms (μ_{B}), magnetic ground state (MGS), the total energy difference between the FM and the AFM configurations ($\Delta E_{\text{FM-AFM}}$), and the critical temperature (T_c) estimated using Equation (3) for Cr_2B_2, Fe_2B_2, and Zr_2B_2.

	Cr_2B_2		Fe_2B_2		Zr_2B_2	
	Pmma	*P6/mmm*	*Pmma*	*P6/mmm*	*Pmma*	*P6/mmm*
μ_{tot} (μ_{B}/unit cell)	2.56	0.63	2.69	0.00	0.00	0.00
μ_{TM} (μ_{B}/ion)	1.03	0.31	1.26	2.06/−2.06	0.00	0.00
μ_{B} (μ_{B}/ion)	−0.05	−0.01	−0.05	0.00	0.00	0.00
MGS	FM	FM	FM	AFM	Non-magnetic	Non-magnetic
$\Delta E_{\text{FM-AFM}}$ (meV/unit cell)	−104.38	−0.29	−108.12	46.84	-	-
T_c (K)	403	1	418	181	-	-

4. Summary and Conclusions

In summary, we carried out a comparison between the structural, electronic, magnetic, and transport properties for orthorhombic and hexagonal phases of a selected group of MBenes, Cr_2B_2, Fe_2B_2 and Zr_2B_2, that are usually considered separately in the literature. Although there are several theoretical reports that have predicted the stability of the studied MBenes, to our knowledge, hex-Fe_2B_2 is shown for the first time in this work to be dynamically stable. Experimentally the most studied MBenes are those composed of early TMs, whereas those composed of late TMs (like iron) remain to be synthesized.

We observe that, from an energetic point of view, the larger the atomic weight is, the higher the E_{coh} is, which means that Zr_2B_2 possesses the strongest bonds. However, despite the small difference between the E_{coh} values of the orthorhombic and hexagonal structures of each MBene, both phases can exist theoretically. This assumption is reinforced by the calculation of the corresponding phonon dispersion plots that predict stable behavior of all the MBenes. Furthermore, the metallic character of our pristine MBenes makes them efficient materials for charge transport, and therefore are competitive 2D materials for electronic, sensing, or electrocatalytic purposes. We predicted that, for both symmetries of Cr_2B_2, the contribution of the majority spin states would be larger than that of the minority spin states at the Fermi level, leading to an FM ground state and opening the possibility of their use in information magnetic storage.

This understanding of the properties of the studied materials, together with the acknowledgment that, within the same framework, both orthorhombic and hexagonal phases are feasible and different in their properties, creates the possibility of going further in our research to consider them as potential candidates for sensing and catalytic processes. Although there exists some parallelism with MXenes, MBenes are emerging 2D materials that are expected to have great development potential in the future, due to their diverse stoichiometries and, as a consequence, structural differences and new challenging physical, chemical, and biological properties. The biggest difference for MBenes is that some of them can potentially exist in both orthorhombic and hexagonal phases.

Author Contributions: Conceptualization, I.M.A.-C. and N.G.S.; methodology, I.M.A.-C.; validation, I.M.A.-C.; investigation, I.M.A.-C. and N.G.S.; writing and original draft preparation, I.M.A.-C.; writing, review, and editing, N.G.S.; visualization, I.M.A.-C.; supervision, N.G.S. All authors have read and agreed to the published version of the manuscript.

Funding: This research received no external funding.

Acknowledgments: The use of supercomputers at the Interdisciplinary Centre for Mathematical and Computational Modelling (ICM) at the University of Warsaw is gratefully acknowledged.

Conflicts of Interest: The authors declare no conflict of interest.

References

1. Akopov, G.; Yeung, M.T.; Kaner, R.B. Rediscovering the Crystal Chemistry of Borides. *Adv. Mater.* **2017**, *29*, 1604506. [CrossRef] [PubMed]
2. Ade, M.; Hillebrecht, H. Ternary Borides Cr_2AlB_2, Cr_3AlB_4, and Cr_4AlB_6: The First Members of the Series $(CrB_2)_nCrAl$ with $n = 1, 2, 3$ and a Unifying Concept for Ternary Borides as MAB-Phases. *Inorg. Chem.* **2015**, *54*, 6122–6135. [CrossRef] [PubMed]
3. Carlsson, A.; Rosen, J.; Dahlqvist, M. Theoretical predictions of phase stability for orthorhombic and hexagonal ternary MAB phases. *Phys. Chem. Chem. Phys.* **2022**, *24*, 11249–11258. [CrossRef] [PubMed]
4. Wang, J.; Ye, T.N.; Gong, Y.; Wu, J.; Miao, N.; Tada, T.; Hosono, H. Discovery of hexagonal ternary phase Ti_2InB_2 and its evolution to layered boride TiB. *Nat. Commun.* **2019**, *10*, 2284. [CrossRef]
5. Alameda, L.T.; Moradifar, P.; Metzger, Z.P.; Alem, N.; Schaak, R.E. Topochemical Deintercalation of Al from MoAlB: Stepwise Etching Pathway, Layered Intergrowth Structures, and Two-Dimensional MBene. *J. Am. Chem. Soc.* **2018**, *140*, 8833–8840. [CrossRef]
6. Alameda, L.T.; Lord, R.W.; Barr, J.A.; Moradifar, P.; Metzger, Z.P.; Steimle, B.C.; Holder, C.F.; Alem, N.; Sinnott, S.B.; Schaak, R.E. Multi-Step Topochemical Pathway to Metastable Mo_2AlB_2 and Related Two-Dimensional Nanosheet Heterostructures. *J. Am. Chem. Soc.* **2019**, *141*, 10852–10861. [CrossRef]
7. Zhang, H.; Xiang, H.; zhi Dai, F.; Zhang, Z.; Zhou, Y. First demonstration of possible two-dimensional MBene CrB derived from MAB phase Cr_2AlB_2. *J. Mater. Sci. Technol.* **2018**, *34*, 2022–2026. [CrossRef]
8. Zhang, H.; Dai, F.Z.; Xiang, H.; Wang, X.; Zhang, Z.; Zhou, Y. Phase pure and well crystalline Cr_2AlB_2: A key precursor for two-dimensional CrB. *J. Mater. Sci. Technol.* **2019**, *35*, 1593–1600. [CrossRef]
9. Khazaei, M.; Wang, J.; Estili, M.; Ranjbar, A.; Suehara, S.; Arai, M.; Esfarjani, K.; Yunoki, S. Novel MAB phases and insights into their exfoliation into 2D MBenes. *Nanoscale* **2019**, *11*, 11305–11314. [CrossRef]
10. Barinov, V.A.; Dorofeev, G.A.; Ovechkin, L.V.; Elsukov, E.P.; Ermakov, A.E. Structure and magnetic properties of the α-FeB phase obtained by mechanical working. *Phys. Status Solidi (a)* **1991**, *123*, 527–534. [CrossRef]

11. Zhao, X.; Li, L.; Bao, K.; Zhu, P.; Tao, Q.; Ma, S.; Cui, T. Insight the effect of rigid boron chain substructure on mechanical, magnetic and electrical properties of β-FeB. *J. Alloys Compd.* **2022**, *896*, 162767. [CrossRef]
12. Saldaña, F.I.; Defoy, E.; Janisch, D.; Rousse, G.; Autran, P.O.; Ghoridi, A.; Séné, A.; Baron, M.; Suescun, L.; Godec, Y.L.; et al. Revealing the Elusive Structure and Reactivity of Iron Boride α-FeB. *Inorg. Chem.* **2023**, *62*, 2073–2082. [CrossRef] [PubMed]
13. Gonzalez Szwacki, N.; Sadrzadeh, A.; Yakobson, B.I. B_{80} Fullerene: An Ab Initio Prediction of Geometry, Stability, and Electronic Structure. *Phys. Rev. Lett.* **2007**, *98*, 166804. [CrossRef] [PubMed]
14. Szwacki, N.G. Boron Fullerenes: A First-Principles Study. *Nanoscale Res. Lett.* **2007**, *3*, 49. [CrossRef]
15. Perdew, J.P.; Burke, K.; Ernzerhof, M. Generalized Gradient Approximation Made Simple. *Phys. Rev. Lett.* **1996**, *77*, 3865–3868. [CrossRef]
16. Blöchl, P.E. Projector augmented-wave method. *Phys. Rev. B* **1994**, *50*, 17953–17979. [CrossRef]
17. Giannozzi, P.; Baroni, S.; Bonini, N.; Calandra, M.; Car, R.; Cavazzoni, C.; Ceresoli, D.; Chiarotti, G.L.; Cococcioni, M.; Dabo, I.; et al. QUANTUM ESPRESSO: A modular and open-source software project for quantum simulations of materials. *J. Phys. Condens. Matter* **2009**, *21*, 395502. [CrossRef]
18. Himmetoglu, B.; Janotti, A. Transport properties of $KTaO_3$ from first-principles. *J. Phys. Condens. Matter* **2016**, *28*, 065502. [CrossRef]
19. Henkelman, G.; Arnaldsson, A.; Jónsson, H. A fast and robust algorithm for Bader decomposition of charge density. *Comput. Mater. Sci.* **2006**, *36*, 354–360. [CrossRef]
20. Momma, K.; Izumi, F. VESTA 3 for three-dimensional visualization of crystal, volumetric and morphology data. *J. Appl. Crystallogr.* **2011**, *44*, 1272–1276. [CrossRef]
21. Bo, T.; Liu, P.F.; Xu, J.; Zhang, J.; Chen, Y.; Eriksson, O.; Wang, F.; Wang, B.T. Hexagonal Ti_2B_2 monolayer: A promising anode material offering high rate capability for Li-ion and Na-ion batteries. *Phys. Chem. Chem. Phys.* **2018**, *20*, 22168–22178. [CrossRef] [PubMed]
22. He, Q.; Li, Z.; Xiao, W.; Zhang, C.; Zhao, Y. Computational investigation of 2D 3*d*/4*d* hexagonal transition metal borides for metal-ion batteries. *Electrochim. Acta* **2021**, *384*, 138404. [CrossRef]
23. Yuan, G.; Bo, T.; Qi, X.; Liu, P.F.; Huang, Z.; Wang, B.T. Monolayer Zr_2B_2: A promising two-dimensional anode material for Li-ion batteries. *Appl. Surf. Sci.* **2019**, *480*, 448–453. [CrossRef]
24. Mir, S.H.; Yadav, V.K.; Singh, J.K. Efficient CO_2 Capture and Activation on Novel Two-Dimensional Transition Metal Borides. *ACS Appl. Mater. Interfaces* **2022**, *14*, 29703–29710. [CrossRef]
25. Dou, M.; Li, H.; Yao, Q.; Wang, J.; Liu, Y.; Wu, F. Room-temperature ferromagnetism in two-dimensional transition metal borides: A first-principles investigation. *Phys. Chem. Chem. Phys.* **2021**, *23*, 10615–10620. [CrossRef]
26. Ozdemir, I.; Kadioglu, Y.; Yüksel, Y.; Ümit Akıncı.; Üzengi Aktürk, O.; Aktürk, E.; Ciraci, S. Columnar antiferromagnetic order of a MBene monolayer. *Phys. Rev. B* **2021**, *103*. [CrossRef]
27. Qi, S.; Fan, Y.; Zhao, L.; Li, W.; Zhao, M. Two-dimensional transition metal borides as highly efficient N_2 fixation catalysts. *Appl. Surf. Sci.* **2021**, *536*, 147742. [CrossRef]
28. Zhang, B.; Zhou, J.; Guo, Z.; Peng, Q.; Sun, Z. Two-dimensional chromium boride MBenes with high HER catalytic activity. *Appl. Surf. Sci.* **2020**, *500* 144248. [CrossRef]
29. Shin, H.; Kang, S.; Koo, J.; Lee, H.; Kim, J.; Kwon, Y. Cohesion energetics of carbon allotropes: Quantum Monte Carlo study. *J. Chem. Phys.* **2014**, *140*, 114702. [CrossRef]
30. Xu, T.; Wang, Y.; Xiong, Z.; Wang, Y.; Zhou, Y.; Li, X. A Rising 2D Star: Novel MBenes with Excellent Performance in Energy Conversion and storage. *Nano-Micro Lett.* **2022**, *15*, 6. [CrossRef]
31. Şahin, H.; Cahangirov, S.; Topsakal, M.; Bekaroglu, E.; Akturk, E.; Senger, R.T.; Ciraci, S. Monolayer honeycomb structures of group-IV elements and III-V binary compounds: First-principles calculations. *Phys. Rev. B* **2009**, *80*, 155453. [CrossRef]
32. Fasolino, A.; Los, J.H.; Katsnelson, M.I. Intrinsic ripples in graphene. *Nat. Mater.* **2007**, *6*, 858–861. [CrossRef] [PubMed]
33. Zhang, B.; Zhou, J.; Sun, Z. MBenes: Progress, challenges and future. *J. Mater. Chem. A* **2022**, *10*, 15865–15880. [CrossRef]
34. Guo, Z.; Zhou, J.; Sun, Z. New two-dimensional transition metal borides for Li ion batteries and electrocatalysis. *J. Mater. Chem. A* **2017**, *5*, 23530–23535. [CrossRef]

Disclaimer/Publisher's Note: The statements, opinions and data contained in all publications are solely those of the individual author(s) and contributor(s) and not of MDPI and/or the editor(s). MDPI and/or the editor(s) disclaim responsibility for any injury to people or property resulting from any ideas, methods, instructions or products referred to in the content.

materials MDPI

Article

Superhard and Superconducting Bilayer Borophene

Chengyong Zhong [1], Minglei Sun [2,*], Tariq Altalhi [3] and Boris I. Yakobson [3,4,*]

[1] College of Physics and Electronic Engineering, Chongqing Normal University, Chongqing 401331, China; zhongcy90@126.com
[2] Department of Materials Science and Nanoengineering, Rice University, Houston, TX 77005, USA
[3] Chemistry Department, Taif University, Taif 21974, Saudi Arabia; ta.altalhi@tu.edu.sa
[4] Department of Chemistry, Rice University, Houston, TX 77005, USA
* Correspondence: minglei.sun@rice.edu (M.S.); biy@rice.edu (B.I.Y.)

Abstract: Two-dimensional superconductors, especially the covalent metals such as borophene, have received significant attention due to their new fundamental physics, as well as potential applications. Furthermore, the bilayer borophene has recently ignited interest due to its high stability and versatile properties. Here, the mechanical and superconducting properties of bilayer-δ_6 borophene are explored by means of first-principles computations and anisotropic Migdal–Eliashberg analytics. We find that the coexistence of strong covalent bonds and delocalized metallic bonds endows this structure with remarkable mechanical properties (maximum 2D-Young's modulus of ~570 N/m) and superconductivity with a critical temperature of ~20 K. Moreover, the superconducting critical temperature of this structure can be further boosted to ~46 K by applied strain, which is the highest value known among all borophenes or two-dimensional elemental materials.

Keywords: bilayer borophene; superhard; anisotropic superconductivity; electron–phonon coupling; strain effect; first-principles calculations

1. Introduction

Superconductivity in two-dimensional (2D) materials has attracted perennial and ever-increasing attention over past decades, owing both to fundamental scientific interest and tantalizing applications [1–5]. One of the most important goals in superconductivity research is raising the superconducting transition temperature, T_c. In general, superconducting materials can be classified into two categories: low-temperature superconductors, characterized by critical temperatures (T_c) below 30 K, such as Nb-Ge films [6], and high-temperature superconductors with T_c exceeding 30 K, exemplified by Cu-based oxides [7–9]. Within the framework of the conventional Bardeen–Cooper–Schrieffer (BCS) theory [10], it is reasonably anticipated that metals composed of light elements have a better chance to induce a high T_c, because the Debye temperatures within such metals are usually high enough to trigger a strong phonon-mediated superconducting pairing. More specifically, according to the celebrated McMillan–Allen–Dynes (MAD) formula [11,12]:

$$T_c = \frac{\omega_{log}}{1.20} exp[\frac{-1.04(1+\lambda)}{\lambda - \mu^*(1+0.62\lambda)}] \quad (1)$$

T_c should be elevated by increasing the log-averaged characteristic phonon frequency (ω_{log}) and the electron–phonon coupling (EPC) parameter, λ. The light elemental materials typically have high frequency phonon modes, enlarging ω_{log} and thus increasing T_c. Furthermore, large phonon frequency and EPC potential $V = \frac{\lambda}{N(E_f)}$ induced by strong covalent bonding in light elemental material, together with $N\left(E_f\right)$, and the electronic density of states (DOS) at the Fermi level [13], should result in a higher T_c. In fact, the metals with strong covalent bonding can be grouped as "covalent metals" [13] and their potential to harbor a high T_c in 2D form have been confirmed, e.g., hydrogenated monolayer MgB_2 (67 K) [14], doped graphene (8.1 K~30 K) [15–17], hydrogenated monolayer borophene (32.4 K) [5] and especially 2D metal borides (1.4 K~72 K) [18–22].

Among covalent metals, 2D boron sheets (borophenes) recently came to the fore of superconducting research, motivated by their inherent metallicity, light weight and significant experimental progress on their synthesis [23–25]. Based on the first-principles calculations and MAD formula, Penev et al. [26] reported early on the T_c of the (experimentally synthesized δ_6, β_{12} and χ_3 borophenes) to be in the range of 11.5 K~20.5 K and Gao et al. [27] found them to span 18.7 K–24.7 K. Using a more accurate and sophisticated anisotropic Migdal–Eliashberg (ME) equation, Zhao et al. [28] found the T_c in those borophenes in 26.2 K–33 K. Apart from intrinsic metallicity and light weight, another impetus for searching high-temperature superconductivity in borophenes is their vast polymorphs and superior mechanical properties, which provide a great benefit to potential superconductors, suggesting effective ways to modulate the superconductivity [29].

The monolayer borophenes have been intensively studied; however, the investigations are scarce [30–32] for bilayer or multilayer borophenes, where more intriguing properties and tunability could be manifested, compared to their monolayer counterparts. Very recently, the realization of bilayer borophenes were reported [33,34]. Unfortunately, the first reported bilayer α-borophene is expected to be unstable if peeled off from the metal substrate [35], and the β_{12}-like bilayer synthesized on Cu(111) surface appears complicated due to ambiguity (with more than three hundred atoms in its unit cell and the atomic structure still in debate), preventing an exploration of its properties or further application. It can be anticipated that more stable and versatile bilayer borophenes are still to be unveiled for research.

In this work, we investigate the mechanical and superconducting properties of a bilayer composed of two covalently bonded δ_6 borophene monolayers, "(BL)-δ_6" in Figure 1a–c. The compact atomic structure endows BL-δ_6 with remarkable stability and tantalizing properties compared to its monolayer counterparts. Its computed high Young's modulus of 570 N/m is even higher than 346 N/m for graphene (or near 280 N/m "per layer", for a fair comparison). Based on the anisotropic ME equation, we find its T_c = 20 K and can be boosted to 46 K by strain effect, reaching the highest T_c found among all borophenes or elemental 2D materials known to date. The coexistence of superhard (in basal plane direction) and superconducting properties in one material is rare, making BL-δ_6 a fascinating superconductor for emergent nanoscale devices such as quantum interferometers, superconducting transistors, superconducting qubits, and wear-resistant parts of superconducting devices [3,36,37].

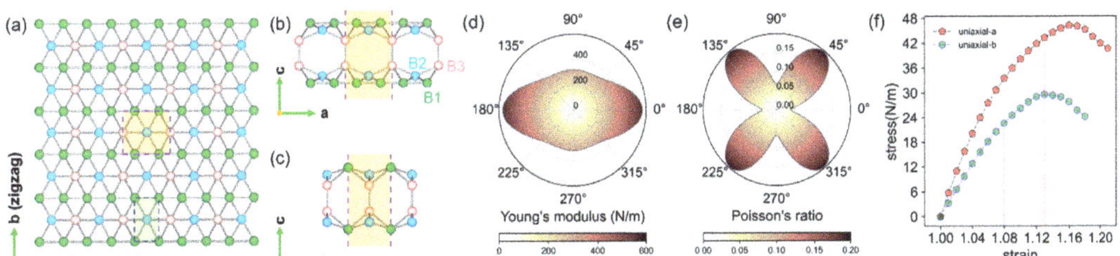

Figure 1. (**a**) Planar and (**b**,**c**) side views of the atomic structure of BL-δ_6. The yellow shaded region is its unit cell, and the green shaded region is the unit cell of δ_6. There are three irreducible boron atoms colored as B1 (green), B2 (blue), and B3 (pink). The direction dependence of (**d**) Young's modulus and (**e**) Poisson's ratio of BL-δ_6. (**f**) Stress–strain curves of BL-δ_6, the vertical blue and red dashed lines denote the fractured stress along the *a* and *b* directions, respectively.

2. Results and Discussion

2.1. Atomic Structure and Mechanical Properties

BL-δ_6 and δ_6 share the same top view (Figure 1a), and BL-δ_6 can be viewed as AB stacking of two δ_6 monolayers bonded via interlayer covalent bonds (side view in Figure 1b,c). The unit cell of BL-δ_6 is a rectangle with lattice constants a = 3.243 Å, b = 2.883 Å (Table 1), which is obtained with a Vienna ab initio Simulation Package [38,39] (computational details can be found in Supplementary Note S1). The total energy of BL-δ_6 is computed to be lower than all experimental synthesized monolayers, due to the interlayer bonding: −6.38 eV/atom, which is 0.171 eV/atom, 0.125 eV/atom and 0.113 eV/atom lower than that of the synthesized δ_6, β_{12} and χ_3, respectively (see Table 1, for atomic structures see Figure S2). Moreover, the slight dynamical instability of δ_6 is also eliminated by the interlayer bonding of BL-δ_6. The energetic stability of BL-δ_6 was also confirmed by an ab initio evolutionary global structure search in our prior work [40] and reported by Zhou et al. [41]. Given that the monolayer δ_6 has been fabricated on a Ag(111) substrate, the thermal stability of BL-δ_6 is also tested on Ag(111) substrate by an AIMD simulation at 500 K, and one can observe that its whole structure is well maintained after 5 ps with a timestep of 1 fs under such a high temperature (Figure S3). Therefore, it is well-expected that BL-δ_6 will be accessed experimentally, supported by its outstanding energetic and thermal stability, and the progress in bilayer borophene synthesis [42,43].

Table 1. The space group, lattice constants (Å), elastic constants (N/m), Young's moduli (N/m), Poisson's ratios, and cohesive energies (eV/atom) of BL-δ_6, δ_6, β_{12}, χ_3.

System	Space Group	a	b	C_{11}	C_{22}	C_{12}	C_{44}	Y_a	Y_b	v_a	v_b	E_c
BL − δ_6	Pmmm	3.243	2.883	570.0	322.0	22.0	146.0	569.4	321.4	0.038	0.068	−6.380
δ_6	Pmmn	1.614	2.874	398.6	171.8	−4.0	93.9	398.6	171.8	−0.01	−0.02	−6.209
δ_6 [24]		1.617	2.865	398.0	170.0	−7.0	94.0	398.0	170.0	−0.04	−0.02	
β_{12}	Pmmm	2.931	5.065	187.0	218.6	36.8	62.7	180.8	211.4	0.168	0.197	−6.255
β_{12} [44]				185.5	210.5	37.0	68.5	179.0	203.1	0.176	0.199	
χ_3	Cmmm	8.407	2.912	207.3	198.5	28.2	60.0	203.3	194.7	0.135	0.142	−6.267
χ_3 [44]				201.0	185.0	21.5	60.5	198.5	182.7	0.116	0.107	
Graphene	P6/mmm	2.468		356.9		62.2	147.3	346.1		0.174		
Graphene [45]		2.470		352.7		60.9	145.9	342.2		0.173		

We obtain the four independent elastic constants for BL-δ_6: C_{11} = 570 N/m, C_{22} = 322 N/m, C_{12} = 22 N/m, and C_{44} = 146 N/m, which apparently satisfy the Born–Huang criteria [46]: $C_{11}C_{22} - C_{12}^2 > 0$ and $C_{44} > 0$, demonstrating the mechanical stability of BL-δ_6. The θ dependence of in-plane Young's modulus Y and Poisson's ratio v of BL-δ_6 are plotted with polar coordinates in Figure 1d,e (computational details can be found in Supplementary Note S2 and Figures S5 and S6). We find that the in-plane Young's moduli and the Poisson's ratios of BL-δ_6 are highly anisotropic. For comparison, the values of elastic constants, Young's moduli, Poisson's ratios of δ_6, β_{12}, χ_3 and graphene are summarized in Table 1. The introduced B3-B3 bonds in BL-δ_6 boost the Y_a and Y_b to 570 N/m and 321 N/m, respectively. Although graphene is well-known as the strongest 2D material, along the armchair direction Y_a = 570 N/m exceeds the Y_a = 364 N/m of graphene, partially due to the strong directional B-B σ-bonds, making it stand out among 2D materials. In contrast, the Y_b = 321 N/m (still comparable to graphene) is lower, due to the much weaker multicenter bonds involved in the zigzag direction.

We find the fracture strain is 13% along the a and 8% along the b direction, according to the strained phonon spectra (Figure S4), before reaching the elastic maximum of 16% and 14% (Figure 1f). Thus, the failure mechanism of BL-δ_6 is phonon instability, for both directions, which is different from that of δ_6 (elastic instability in the zigzag direction and phonon instability in the armchair direction) [47]. The fracture strengths of BL-δ_6 are 43 and 23 N/m along the a and b directions, much above that of δ_6 (20.26 and 12.98 N/m) [47], black phosphorene (9.99 and 4.44 N/m) [48], and even graphene (40.41 and 36.74 N/m) [49] along the zigzag direction. Considering the ultrahigh Young's moduli and large fracture strength, one can judge the BL-δ_6 being in-plane superhard.

2.2. Electronic Properties

The electronic structure of BL-δ_6 is summarized in Figure 2. Three bands cross the Fermi level and form the Fermi surface, of which one forms a small electron pocket centered around the corner of the first BZ and the other two contribute to the rest of Fermi surface (Figure 2b). The atomic- and orbital-resolved band structures suggest that the main contribution near the Fermi level is from the p_y orbital of B1/B3 atoms and the p_z orbital of all B atoms (Figure 2f,g, more details in Figure S7), which is also confirmed by the charge distribution in the states within $\pm$0.3 eV from the Fermi level (Figure 2c–e). Using Wannier90 (v 3.1.0) code [50], the Wannier-interpolated band structures are also provided and show excellent agreement with those by first-principles calculations (Figure 2a), which lays a clear ground for subsequent anisotropic superconductivity calculations, as implemented in electron–phonon Wannier (EPW v 5.4) code [51].

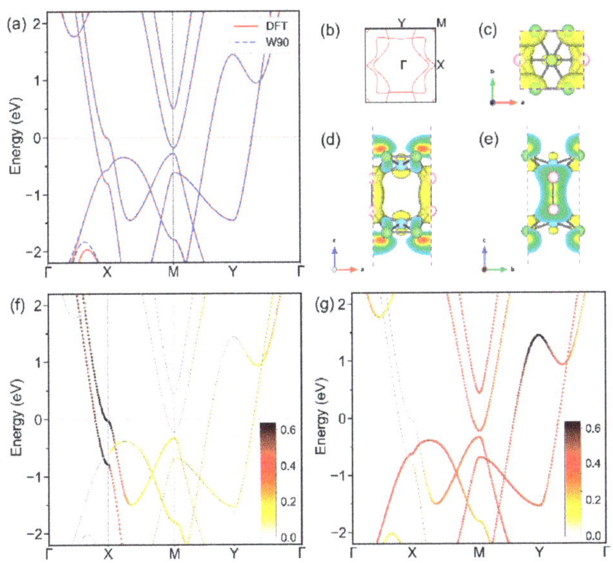

Figure 2. (**a**) Electronic band structures of BL-δ_6 by DFT (red solid lines) and Wannier90 (blue dashed lines). (**b**) Fermi surface, (**c–e**) top and side views of the charge density corresponding to the energy range shaded green in (**a**), and electronic band structures weighted by the p_y orbital of B1 and B3 atoms (**f**) and the p_z orbital contributions of all boron atoms (**g**) of BL-δ_6.

2.3. Isotropic and Anisotropic Superconducting Properties

In the phonon spectrum of BL-δ_6 (Figure 3a), we find no negative frequencies, demonstrating its dynamic stability. This is in stark contrast to δ_6, a layer whose phonon dispersion displays small imaginary frequency near the Γ point [24]. Mechanically intuitive, the BL-δ_6 is stabilized by the covalent B3-B3 bonds. The (ν) mode- and (q) momentum-resolved EPC $\lambda_{q\nu}$ is also given, calculated by the following equation:

$$\lambda_{\mathrm{q\nu}} = \frac{2}{\hbar N(f) N_k} \sum\nolimits_{nmk} \frac{1}{\omega_{q\nu}} \left| g_{k,q\nu}^{nm} \right|^2 \delta(\epsilon_k^n)\delta(\epsilon_{k+q}^m), \tag{2}$$

where N_{k} represents the total number of k points in the k space, $g_{k,q\nu}^{nm}$ is an EPC matrix element, n/m and ν represent the indices of electronic bands and phonon mode, ϵ_k^n and ϵ_{k+q}^m

are the eigenvalues with respect to the Fermi level, and ω_{qv} is the phonon frequency. We found three modes (B_{2u} and two A_g) have comparatively large EPC near the Γ point (left panel of Figure 3a). In the B_{2u} shear mode at 39.3 meV, the B1 and B2 atoms move in one direction and the B3 atoms move in the opposite (Figure 3b). The first A_g mode at 102.6 meV corresponds to a bond stretching contributed by B3 atoms (Figure 3c). The second A_g mode at 168.7 meV is also a stretch mode, but predominantly contributed by B1 atoms (Figure 3d). The vibrational contribution of these three modes can also be inferred from the phonon density of states (right panel of Figure 3a). The Eliashberg spectral function $\alpha^2F(\omega)$ is a key quantity, which determines the total EPC λ by the following equation:

$$\lambda = \sum_{qv} \lambda_{qv} = 2\int \frac{\alpha^2F(\omega)}{\omega} d\omega \tag{3}$$

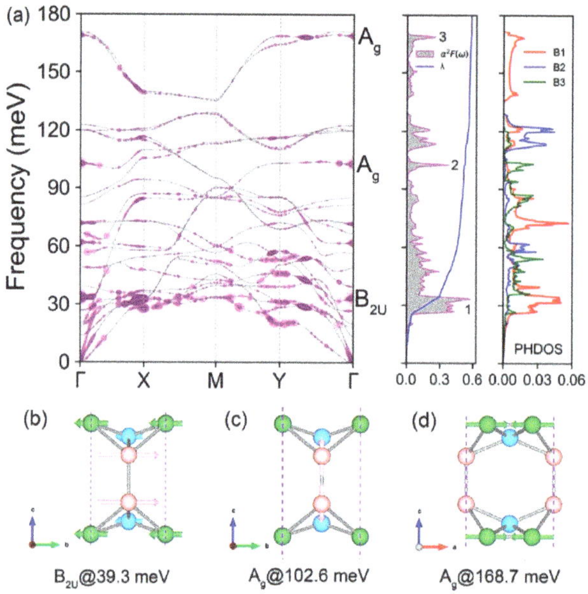

Figure 3. (**a**) From left to right: phonon band structure of BL-δ_6 weighted by EPC λ_{qv} with purple circles, isotropic Eliashberg function α^2F and EPC $\lambda(\omega)$, and phonon density of states contributed by different kinds of boron atoms. (**b–d**) Vibrational modes of the B_{2u} mode and the two A_g modes at Γ, as labeled in (**a**).

Accordingly, the $\alpha^2F(\omega)$ and cumulative EPC $\lambda(\omega)$ are calculated (middle panel of Figure 3a). We observe three sharp peaks mainly contributed by the three phonon modes analyzed above. Correspondingly, the EPC strength λ calculated by Equation (3) is 0.59 and the T_c of BL-δ_6 within the MAD approximation calculated by Equation (1) is 10.1 K.

Akin to structure and mechanics anisotropy, the Fermi surface of BL-δ_6, formed by multiple bands with different orbital contributions (p_y and p_z orbitals, Figure 2f,g) is also significantly anisotropic. For such a system with a complicated Fermi surface, using the anisotropic ME equation is essential, in order to obtain an anisotropic EPC, and an accurate Tc, compared with the isotropic superconductivity calculated from the MAD formula using an isotropic EPC [28]. As shown in Figure 4a,b, the variation in momentum-dependent EPC parameter λ_k and the superconducting gap Δ_k at 10 K displays similar anisotropy, i.e., the maximum along the $\Gamma - X$ direction and the minimum along the $\Gamma - Y$ direction. We find the superconducting gap ratio $\Delta^{aniso} = (\Delta^{max} - \Delta^{min})/\Delta^{ave} = (2.499 - 0.444)/1.470 \approx 140\%$, a measure of its strong anisotropy at the Fermi surface, also signifying that the anisotropic

ME formula is indispensable for predicting T_c. Figure 4c shows the evolution of the superconducting gap as a function of temperature, based on the ME equations solved in either isotropic or fully anisotropic approximations. The T_c is identified as the lowest temperature at which the vanishing gap is observed. Under the isotropic approximation T_c = 12 K, comparable to the value (10 K) obtained by the MAD formula but is much lower than the value calculated with the fully anisotropic approximation (20 K), further attesting to the anisotropic superconducting nature of BL-δ_6, omitted previously [32].

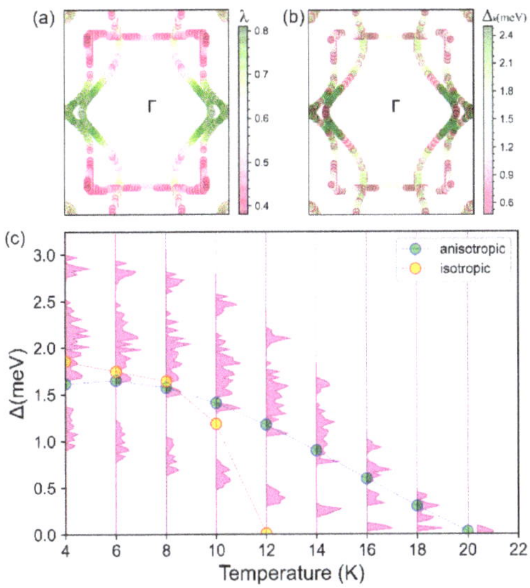

Figure 4. (**a**) Momentum-dependent EPC λ_k across the full BZ and (**b**) momentum-dependent superconducting band gap Δ_k at 10 K and projected onto the Fermi surface. (**c**) Variation in the superconducting gap Δ_k with temperature, calculated by solving the ME equations in the isotropic approximation (yellow dots and dashed line interpolated) and with the fully anisotropic solution, where the purple shadowed regions indicate the magnitude distribution of the Δ_k and the light green dots connected with the dashed line represents the average value of the entire anisotropic Δ_k.

The mechanical robustness of BL-δ_6 permits a consideration of whether the tensile strain could enhance the T_c, motivated mainly by two reasons. First, according to Equation (2), it can be seen that EPC λ_{qv} is inversely proportional to the phonon frequency ω_{qv}, which would be helpful to lower, perhaps by tension, weakening the atomic force constants and thus softening the phonon modes. Second, the atomic orbitals overlap is reduced so that the electronic bands become less dispersive, enlarging the $N(E_f)$, which provides more electrons susceptible to pairing interactions, mediated by dynamic phonons, to certainly contribute to the T_c rise.

To examine the effect of strain on the BL-δ_6 superconductivity, the evolution of EPC λ, the log and maximum frequencies ω_{log} and ω_{max}, as well as T_c, are all calculated within three approximations (i.e., MAD, isotropic ME, anisotropic ME), as a function of tensile strain along the *a* direction, and presented in Figure 5a,b (for details see Figures S8–S11). As expected, the increment in T_c is accompanied by a decrease in frequency-dependent $\omega_{log}/\omega_{max}$ and an increase in EPC λ (Figure 5b). The phonon spectrum under a 13% tension along *a* (near fracture strain) weighted by λ_{qv}, the Eliashberg spectral function $\alpha^2F(\omega)$ and EPC $\lambda(\omega)$ are shown in Figure 5c,d. The red shift of phonon frequency can be clearly observed and the enhanced EPC λ_{qv} due to the phonon softening can be seen in all frequency ranges (Figure 5d). In addition, the contribution of low-frequency part becomes more dominant. In particular, four largely softened phonon Kohn anomalies with considerable EPCs around 15 meV appear on the path of X-M-Y. Eventually, these combined effects lead to a significant increase, by strain, in λ from 0.59 to 1.33 and T_c from 20 K to 46 K.

It should be emphasized that 46 K is the highest T_c among all borophenes and elemental 2D materials (Table 2). It should be noted that for bilayer or multilayer materials with positive Poisson ratios, such as BL-δ_6, the tensile strain would lead to vertical shrinking (Figure 5a), suggesting that the vertical pressure may also be an effective way to enhance the T_c of BL-δ_6, as the tensile strain does.

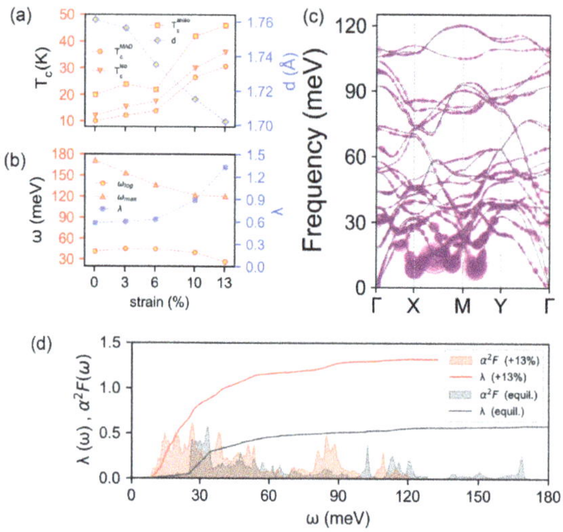

Figure 5. (**a**) Evolution with tensile strain along the *a* direction, of the T_c at three approximations (MAD, circles; isotropic ME, triangles; anisotropic ME, squares) and the distance between δ_6 planes measured as B3-B3 bond length *d*. (**b**) Evolution of EPC λ (solid crosses), the log-weighted frequency ω_{log}, (circles) and maximum phonon frequency ω_{max} (triangles) versus the strain. (**c**) Phonon band structure weighted by EPC ω_{qv} under 13% strain. (**d**) Isotropic Eliashberg $\alpha_2 F$ and EPC λ with 13% tensile strain along the *a* direction. Data with no strain are shown in gray, for comparison.

Table 2. The EPC λ, T_c calculated with MAD formula and anisotropic ME equation of δ_6, β_{12}, χ_3, for Li-doped graphene [15] and BL-δ_6.

System	λ	T_c^{MAD} ($K, \mu^* = 0.1$)	T_c^{aniso} ($K, \mu^* = 0.1$)
δ_6 [26]	1.10	20.5	
δ_6 [28]	0.82		27.0
β_{12} [26]	0.80	16.1	
β_{12} [27]	0.89	18.7	
β_{12} [28]	1.01		33.0
χ_3 [26]	0.60	11.5	
χ_3 [27]	0.95	24.7	
χ_3 [28]	0.79		26.2
Graphene (Li deposition) [15]	0.61	8.1	
BL − δ_6	0.59	10.1	20
BL − δ_6 [32]	0.61	11.9	
BL − δ_6 (13%)	1.33	30.78	46

Before concluding, a few remarks need to be addressed. Superhard materials are usually semiconductors with large bandgaps or insulators, because the majority of electrons are bound into covalent bonds, depleting any excess electrons from electron transport. For instance, diamond is an insulator and the hardest material known. It may seem that metallicity and superhardness cannot coexist in one covalent material. However, this dilemma can be resolved in 2D borophenes, because the electron deficiency nature of boron atom enables strong localized covalent bonds and delocalized multicenter metal bonds in one system, which endows the system with superb mechanical performance and excellent metallicity, as evidenced by the high Young's moduli and metallicity in BL-δ_6.

It is worth mentioning that research on BL borophenes is still in its infancy and has just been ignited by two recent independent experimental investigations. We would also like to point out that in addition to the striking mechanical properties and superconductivity studied in this work, BL-δ_6 is expected to possess other interesting properties, such as anisotropic plasmonics [52] and ultrahigh thermal conductivity [53], awaiting more research. On the other hand, the excellent mechanical stability is also beneficial for BL-δ_6 transfer to or even growth on some inert substrates [54], for convenience of characterization. Our results also show that the interlayer bonds, while strengthening the bilayer, do not destroy the σ-bond resonance nor the conducting π-bonds [55], adding a fine tuning-knob to the properties. This also highlights the potential of covalent metals in the quest for high T_c superconductors.

In our investigation, the tensile strain is applied to enhance the superconducting temperature of bilayer borophene. In addition to strain, the phenomenon of disorder, particularly correlated disorder, can also significantly enhance the superconductivity in a material [56,57]. In bilayer borophene, manipulating disorder could pave new paths to enhance superconducting performance. On the other hand, a caveat common for all $\leq$2D materials, due to the Mermin–Wagner theorem, must be mentioned, since it restricts the stability, both merely structural and of the superconducting phase. Nevertheless, in any BL borophene realization, the sample finite size (and support by a 3D-substrate) should mitigate these concerns, although quantifying the size limitation is far beyond the scope of the present study.

3. Conclusions

In conclusion, we comprehensively investigated mechanical and superconducting properties of a bilayer borophene (BL-δ_6), which can be viewed as AB stacking, with interlayer covalent bonding, of earlier realized δ_6 borophene. The original good stability of δ_6 and the introduced covalent bond endows BL-δ_6 with very prominent energy—thermodynamic, thermal, and mechanical stability, suggesting that it has a high chance of occurrence in experiments. The strong directional σ B-B bonds and very compact atomic configuration give BL-δ_6 a 2D Young's modulus along the *a* direction (569.4 N/m) higher than graphene (346.1 N/m), while ensuring that BL-δ_6 can sustain an ultimate 13% and 8% tensile strain along the *a* and *b* directions, respectively. Furthermore, according to a fully anisotropic solution of the ME equation, we predict BL-δ_6 is a conventional phonon-mediated 2D superconductor with T_c as high as 20 K. We also highlighted the effects of tensile strain on EPCs and found that the T_c can be boosted to 46 K at a 13% tensile strain along the *a* direction. To the best of our knowledge, it is the highest value currently known for borophene or elemental 2D materials. Our findings also consolidated the justification that covalent metals, such as borophene, should benefit the search for high T_c superconductors. The concurrent superior mechanical performance and excellent superconductivity is scarce; thus, promising BL-δ_6 many potential applications, such as quantum interferometers, superconducting transistors, superconducting qubits, and even wear-resistant parts for superconducting devices.

Supplementary Materials: The following supporting information can be downloaded at: https://www.mdpi.com/article/10.3390/ma17091967/s1. Figure S1: The evolution of T_c with different *k* and *q* meshes. Figure S2: The atomic structures of (a) δ_6, (b) β_{12} and (c) χ_3. Figure S3: (a) The total potential energy fluctuation of BL-δ_6 during AIMD simulation at 500 K. (b) The top view and (c) side view of the atomic configuration of BL-δ_6 on Ag (111) surface after 5 ps of AIMD simulation at 500 K. Figure S4: The phonon spectra of strains under (a) 13% and (b) 14% along a axis, (c) 8% and (d) 9% along b axis. Figure S5: The strain energy per area under different kinds of strains for BL-δ_6. Figure S6: The strain energy per area under different kinds of strains for (a) δ_6, (b) β_{12}, (c) χ_3 and (d) graphene. Figure S7: The atomic and orbital resolved band structures of BL-δ_6. Figure S8: (a) The electronic band properties of BL-δ_6 under 3% tensile strain along a direction, red line and blue dashed line denote the results calculated from DFT and Wannier90, respectively. (b) Phonon band structure of BL-δ_6 weighted by EPC λ_{qv} with purple circles, isotropic Eliashberg function α^2F and EPC $\lambda(\omega)$, PHDOS from different kinds of boron atoms' contribution. (c) Evolution of the superconducting gap Δ_k as a function of temperature, calculated by solving the ME equations in the isotropic approximation (yellow dots and dashed line interpolation) and with a fully anisotropic solution where the purple shadowed regions indicate the magnitude distribution of the Δ_k and the light green dots connected with dashed line represents the average value of the entire anisotropic Δ_k. Figure S9: The results of BL-δ_6 under 6% tensile strain along a direction. The meaning of (a–c) are the same as Figure S8. Figure S10: The results of BL-δ_6 under 10% tensile strain along a direction. The meaning of (a–c) are the same as Figure S8. Figure S11: The results of BL-δ_6 under 13% tensile strain along a direction. The meaning of (a–c) are the same as Figure S8. Refs. [38,39,50,51,58–65] are cited in the Supplementary Materials.

Author Contributions: Conceptualization, C.Z. and B.I.Y.; Methodology, M.S.; Investigation, C.Z. and M.S.; Resources, C.Z., T.A. and B.I.Y.; Writing—original draft, C.Z. and M.S.; Writing—review & editing, T.A. and B.I.Y. All authors have read and agreed to the published version of the manuscript.

Funding: C.Z. acknowledges the financial support from the Scientific and Technology Research Program of Chongqing Municipal Education Commission (KJQN202300515) and the Foundation of Chongqing Normal University (23XLB001). T.A. and B.I.Y. acknowledge the Taif University Research Support Project (TURSPHC2024/1, Saudi Arabia). Work at Rice was supported by the US Office of Naval Research (N00014-22-1-2753).

Institutional Review Board Statement: Not applicable.

Informed Consent Statement: Not applicable.

Data Availability Statement: Data are contained within the article.

Conflicts of Interest: The authors declare no conflicts of interest.

References

1. Saito, Y.; Nojima, T.; Iwasa, Y. Highly crystalline 2D superconductors. *Nat. Rev. Mater.* **2016**, *2*, 16094. [CrossRef]
2. Li, W.; Huang, J.; Li, X.; Zhao, S.; Lu, J.; Han, Z.V.; Wang, H. Recent progresses in two-dimensional Ising superconductivity. *Mater. Today Phys.* **2021**, *21*, 100504. [CrossRef]
3. Qiu, D.; Gong, C.; Wang, S.; Zhang, M.; Yang, C.; Wang, X.; Xiong, J. Recent Advances in 2D Superconductors. *Adv. Mater.* **2021**, *33*, e2006124. [CrossRef] [PubMed]
4. Lilia, B.; Hennig, R.; Hirschfeld, P.; Profeta, G.; Sanna, A.; Zurek, E.; Pickett, W.E.; Amsler, M.; Dias, R.; Eremets, M.I.; et al. The 2021 room-temperature superconductivity roadmap. *J. Phys. Condens. Matter* **2022**, *34*, 183002.
5. Zhong, C.; Li, X.; Yu, P. Strain-tunable Dirac semimetal phase transition and emergent superconductivity in a borophane. *Commun. Phys.* **2024**, *7*, 38. [CrossRef]
6. Gavaler, J.R. Superconductivity in Nb–Ge films above 22 K. *Appl. Phys. Lett.* **1973**, *23*, 480–482. [CrossRef]
7. Wu, M.K.; Ashburn, J.R.; Torng, C.J.; Hor, P.H.; Meng, R.L.; Gao, L.; Huang, Z.J.; Wang, Y.Q.; Chu, C.W. Superconductivity at 93 K in a new mixed-phase Y-Ba-Cu-O compound system at ambient pressure. *Phys. Rev. Lett.* **1987**, *58*, 908–910. [CrossRef] [PubMed]
8. Pęczkowski, P.; Zachariasz, P.; Kowalik, M.; Tokarz, W.; Kumar Naik, S.P.; Żukrowski, J.; Jastrzębski, C.; Dadiel, L.J.; Tabiś, W.; Gondek, Ł. Iron diffusivity into superconducting $YBa_2Cu_3O_{7-\delta}$ at oxygen-assisted sintering: Structural, magnetic, and transport properties. *J. Eur. Ceram. Soc.* **2021**, *41*, 7085–7097. [CrossRef]
9. Peczkowski, P.; Kowalik, M.; Zachariasz, P.; Jastrzebski, C.; Jaegermann, Z.; Szterner, P.; Woch, W.M.; Szczytko, J. Synthesis and Physicochemical Properties of Nd-, Sm-, Eu-Based Cuprate High-Temperature Superconductors. *Phys. Status Solidi A* **2018**, *215*, 1700888. [CrossRef]
10. Bardeen, J.; Cooper, L.N.; Schrieffer, J.R. Theory of Superconductivity. *Phys. Rev.* **1957**, *108*, 1175–1204. [CrossRef]
11. McMillan, W.L. Transition Temperature of Strong-Coupled Superconductors. *Phys. Rev.* **1968**, *167*, 331–344. [CrossRef]
12. Allen, P.B.; Dynes, R.C. Transition temperature of strong-coupled superconductors reanalyzed. *Phys. Rev. B* **1975**, *12*, 905–922. [CrossRef]
13. Blase, X.; Bustarret, E.; Chapelier, C.; Klein, T.; Marcenat, C. Superconducting group-IV semiconductors. *Nat. Mater.* **2009**, *8*, 375–382. [CrossRef]
14. Bekaert, J.; Petrov, M.; Aperis, A.; Oppeneer, P.M.; Milosevic, M.V. Hydrogen-Induced High-Temperature Superconductivity in Two-Dimensional Materials: The Example of Hydrogenated Monolayer MgB_2. *Phys. Rev. Lett.* **2019**, *123*, 077001. [CrossRef] [PubMed]
15. Profeta, G.; Calandra, M.; Mauri, F. Phonon-mediated superconductivity in graphene by lithium deposition. *Nat. Phys.* **2012**, *8*, 131–134. [CrossRef]

16. Lu, H.-Y.; Yang, Y.; Hao, L.; Wang, W.-S.; Geng, L.; Zheng, M.; Li, Y.; Jiao, N.; Zhang, P.; Ting, C.S. Phonon-mediated superconductivity in aluminum-deposited graphene AlC_8. *Phys. Rev. B* **2020**, *101*, 214514. [CrossRef]
17. Si, C.; Liu, Z.; Duan, W.; Liu, F. First-principles calculations on the effect of doping and biaxial tensile strain on electron-phonon coupling in graphene. *Phys. Rev. Lett.* **2013**, *111*, 196802. [CrossRef]
18. Zhao, Y.; Lian, C.; Zeng, S.; Dai, Z.; Meng, S.; Ni, J. Two-gap and three-gap superconductivity in AlB_2-based films. *Phys. Rev. B* **2019**, *100*, 094516. [CrossRef]
19. Bo, T.; Liu, P.-F.; Yan, L.; Wang, B.-T. Electron-phonon coupling superconductivity in two-dimensional orthorhombic MB_6 (M = Mg, Ca, Ti, Y) and hexagonal MB_6 (M = Mg, Ca, Sc, Ti). *Phys. Rev. Mater.* **2020**, *4*, 114802. [CrossRef]
20. Zhao, Y.; Lian, C.; Zeng, S.; Dai, Z.; Meng, S.; Ni, J. MgB_4 trilayer film: A four-gap superconductor. *Phys. Rev. B* **2020**, *101*, 104507. [CrossRef]
21. Sevik, C.; Bekaert, J.; Petrov, M.; Milošević, M.V. High-temperature multigap superconductivity in two-dimensional metal borides. *Phys. Rev. Mater.* **2022**, *6*, 024803. [CrossRef]
22. Wang, Z.; Zeng, S.; Zhao, Y.; Wang, X.; Ni, J. Three-gap superconductivity in two-dimensional InB_2/InB_4 films. *Phys. Rev. B* **2021**, *104*, 174519. [CrossRef]
23. Kaneti, Y.V.; Benu, D.P.; Xu, X.; Yuliarto, B.; Yamauchi, Y.; Golberg, D. Borophene: Two-dimensional Boron Monolayer: Synthesis, Properties, and Potential Applications. *Chem. Rev.* **2022**, *122*, 1000–1051. [CrossRef] [PubMed]
24. Mannix, A.J.; Zhou, X.-F.; Kiraly, B.; Wood, J.D.; Alducin, D.; Myers, B.D.; Liu, X.; Fisher, B.L.; Santiago, U.; Guest, J.R.; et al. Synthesis of borophenes: Anisotropic, two-dimensional boron polymorphs. *Science* **2015**, *350*, 1513–1516. [CrossRef]
25. Feng, B.; Zhang, J.; Zhong, Q.; Li, W.; Li, S.; Li, H.; Cheng, P.; Meng, S.; Chen, L.; Wu, K. Experimental realization of two-dimensional boron sheets. *Nat. Chem.* **2016**, *8*, 563–568. [CrossRef]
26. Penev, E.S.; Kutana, A.; Yakobson, B.I. Can Two-Dimensional Boron Superconduct? *Nano Lett.* **2016**, *16*, 2522–2526. [CrossRef] [PubMed]
27. Gao, M.; Li, Q.-Z.; Yan, X.-W.; Wang, J. Prediction of phonon-mediated superconductivity in borophene. *Phys. Rev. B* **2017**, *95*, 024505. [CrossRef]
28. Zhao, Y.; Zeng, S.; Lian, C.; Dai, Z.; Meng, S.; Ni, J. Multigap anisotropic superconductivity in borophenes. *Phys. Rev. B* **2018**, *98*, 134514. [CrossRef]
29. Zhang, Z.; Penev, E.S.; Yakobson, B.I. Two-dimensional materials: Polyphony in B flat. *Nat. Chem.* **2016**, *8*, 525–527. [CrossRef]
30. Gao, N.; Wu, X.; Jiang, X.; Bai, Y.; Zhao, J. Structure and stability of bilayer borophene: The roles of hexagonal holes and interlayer bonding. *FlatChem* **2017**, *7*, 48–54. [CrossRef]
31. Zhong, H.; Huang, K.; Yu, G.; Yuan, S. Electronic and mechanical properties of few-layer borophene. *Phys. Rev. B* **2018**, *98*, 054104. [CrossRef]
32. Yan, L.; Ku, R.; Zou, J.; Zhou, L.; Zhao, J.; Jiang, X.; Wang, B.-T. Prediction of superconductivity in bilayer borophenes. *RSC Adv.* **2021**, *11*, 40220–40227. [CrossRef] [PubMed]
33. Liu, X.; Li, Q.; Ruan, Q.; Rahn, M.S.; Yakobson, B.I.; Hersam, M.C. Borophene synthesis beyond the single-atomic-layer limit. *Nat. Mater.* **2022**, *21*, 35–40. [CrossRef]
34. Chen, C.; Lv, H.; Zhang, P.; Zhuo, Z.; Wang, Y.; Ma, C.; Li, W.; Wang, X.; Feng, B.; Cheng, P.; et al. Synthesis of bilayer borophene. *Nat. Chem.* **2022**, *14*, 25–31. [CrossRef] [PubMed]
35. Ma, Y.-Y.; Zhao, X.-Y.; Zan, W.; Mu, Y.; Zhang, Z.; Li, S.-D. Prediction of freestanding semiconducting bilayer borophenes. *Nano Res.* **2022**, *15*, 5752–5757. [CrossRef]
36. Liu, X.; Hersam, M.C. 2D materials for quantum information science. *Nat. Rev. Mater.* **2019**, *4*, 669–684. [CrossRef]
37. Liu, P.; Lei, B.; Chen, X.; Wang, L.; Wang, X. Superior carrier tuning in ultrathin superconducting materials by electric-field gating. *Nat. Rev. Phys.* **2022**, *4*, 336–352. [CrossRef]
38. Kresse, G.; Furthmüller, J. Efficiency of ab-initio total energy calculations for metals and semiconductors using a plane-wave basis set. *Comp. Mater. Sci.* **1996**, *6*, 15–50. [CrossRef]
39. Kresse, G.; Furthmüller, J. Efficient iterative schemes for ab initio total-energy calculations using a plane-wave basis set. *Phys. Rev. B* **1996**, *54*, 11169–11186. [CrossRef] [PubMed]
40. Sun, M.; Luo, Y.; Yan, Y.; Schwingenschlögl, U. Ultrahigh Carrier Mobility in the Two-Dimensional Semiconductors B_8Si_4, B_8Ge_4, and B_8Sn_4. *Chem. Mater.* **2021**, *33*, 6475–6483. [CrossRef]
41. Zhou, X.-F.; Dong, X.; Oganov, A.R.; Zhu, Q.; Tian, Y.; Wang, H.-T. Semimetallic Two-Dimensional Boron Allotrope with Massless Dirac Fermions. *Phys. Rev. Lett.* **2014**, *112*, 085502. [CrossRef]
42. Ebrahimi, M. The birth of bilayer borophene. *Nat. Chem.* **2022**, *14*, 3–4. [CrossRef] [PubMed]
43. Božović, I. Doubling down on borophene electronics. *Nat. Mater.* **2022**, *21*, 11–12. [CrossRef] [PubMed]
44. Wang, Z.; Lü, T.-Y.; Wang, H.-Q.; Feng, Y.P.; Zheng, J.-C. High anisotropy of fully hydrogenated borophene. *Phys. Chem. Chem. Phys.* **2016**, *18*, 31424–31430. [CrossRef] [PubMed]
45. Andrew, R.C.; Mapasha, R.E.; Ukpong, A.M.; Chetty, N. Mechanical properties of graphene and boronitrene. *Phys. Rev. B* **2012**, *85*, 125428. [CrossRef]
46. Nye, J.F. *Physical Properties of Crystals*; Clarendon Press: Oxford, UK, 1985.
47. Wang, H.; Li, Q.; Gao, Y.; Miao, F.; Zhou, X.-F.; Wan, X.G. Strain effects on borophene: Ideal strength, negative Possion's ratio and phonon instability. *New J. Phys.* **2016**, *18*, 073016. [CrossRef]

48. Wei, Q.; Peng, X. Superior mechanical flexibility of phosphorene and few-layer black phosphorus. *Appl. Phys. Lett.* **2014**, *104*, 251915. [CrossRef]
49. Liu, F.; Ming, P.; Li, J. Ab initio calculation of ideal strength and phonon instability of graphene under tension. *Phys. Rev. B* **2007**, *76*, 064120. [CrossRef]
50. Pizzi, G.; Vitale, V.; Arita, R.; Blügel, S.; Freimuth, F.; Géranton, G.; Gibertini, M.; Gresch, D.; Johnson, C.; Koretsune, T.; et al. Wannier90 as a community code: New features and applications. *J. Phys. Condens. Matter.* **2020**, *32*, 165902. [CrossRef]
51. Noffsinger, J.; Giustino, F.; Malone, B.D.; Park, C.-H.; Louie, S.G.; Cohen, M.L. EPW: A program for calculating the electron–phonon coupling using maximally localized Wannier functions. *Comput. Phys. Commun.* **2010**, *181*, 2140–2148. [CrossRef]
52. Huang, Y.; Shirodkar, S.N.; Yakobson, B.I. Two-Dimensional Boron Polymorphs for Visible Range Plasmonics: A First-Principles Exploration. *J. Am. Chem. Soc.* **2017**, *139*, 17181–17185. [CrossRef]
53. Xiao, H.; Cao, W.; Ouyang, T.; Guo, S.; He, C.; Zhong, J. Lattice thermal conductivity of borophene from first principle calculation. *Sci. Rep.* **2017**, *7*, 45986. [CrossRef] [PubMed]
54. Ruan, Q.; Wang, L.; Bets, K.V.; Yakobson, B.I. Step-Edge Epitaxy for Borophene Growth on Insulators. *ACS Nano* **2021**, *15*, 18347–18353. [CrossRef] [PubMed]
55. Qiu, L.; Zhang, X.; Kong, X.; Mitchell, I.; Yan, T.; Kim, S.Y.; Yakobson, B.I.; Ding, F. Theory of sigma bond resonance in flat boron materials. *Nat. Comm.* **2023**, *14*, 1804. [CrossRef] [PubMed]
56. Neverov, V.D.; Lukyanov, A.E.; Krasavin, A.V.; Vagov, A.; Croitoru, M.D. Correlated disorder as a way towards robust superconductivity. *Commun. Phys.* **2022**, *5*, 177. [CrossRef]
57. Ghosal, A.; Randeria, M.; Trivedi, N. Inhomogeneous pairing in highly disordered s-wave superconductors. *Phys. Rev. B* **2001**, *65*, 014501. [CrossRef]
58. Perdew, J.P.; Chevary, J.A.; Vosko, S.H.; Jackson, K.A.; Pederson, M.R.; Singh, D.J.; Fiolhais, C. Atoms, molecules, solids, and surfaces: Applications of the generalized gradient approximation for exchange and correlation. *Phys. Rev. B* **1992**, *46*, 6671–6687. [CrossRef]
59. Blöchl, P.E. Projector augmented-wave method. *Phys. Rev. B* **1994**, *50*, 17953–17979. [CrossRef] [PubMed]
60. Evans, D.J.; Holian, B.L. The Nose–Hoover thermostat. *J. Chem. Phys.* **1985**, *83*, 4069–4074. [CrossRef]
61. Nosé, S. A unified formulation of the constant temperature molecular dynamics methods. *J. Chem. Phys.* **1984**, *81*, 511–519. [CrossRef]
62. Giannozzi, P.; Andreussi, O.; Brumme, T.; Bunau, O.; Nardelli, M.B.; Calandra, M.; Car, R.; Cavazzoni, C.; Ceresoli, D.; Cococcioni, M.; et al. Advanced capabilities for materials modelling with Quantum ESPRESSO. *J. Phys. Condens. Matter* **2017**, *29*, 465901. [CrossRef] [PubMed]
63. Baroni, S.; de Gironcoli, S.; Dal Corso, A.; Giannozzi, P. Phonons and related crystal properties from density-functional perturbation theory. *Rev. Mod. Phys.* **2001**, *73*, 515–562. [CrossRef]
64. Zhong, C.; Wu, W.; He, J.; Ding, G.; Liu, Y.; Li, D.; Yang, S.A.; Zhang, G. Two-dimensional honeycomb borophene oxide: Strong anisotropy and nodal loop transformation. *Nanoscale* **2019**, *11*, 2468–2475. [CrossRef] [PubMed]
65. Cadelano, E.; Palla, P.L.; Giordano, S.; Colombo, L. Elastic properties of hydrogenated graphene. *Phys. Rev. B* **2010**, *82*, 235414. [CrossRef]

Disclaimer/Publisher's Note: The statements, opinions and data contained in all publications are solely those of the individual author(s) and contributor(s) and not of MDPI and/or the editor(s). MDPI and/or the editor(s) disclaim responsibility for any injury to people or property resulting from any ideas, methods, instructions or products referred to in the content.

materials

MDPI

Article

Adsorption of Atomic Hydrogen on Hydrogen Boride Sheets Studied by Photoelectron Spectroscopy

Heming Yin [1], Jingmin Tang [1,*], Kazuki Yamaguchi [1], Haruto Sakurai [1], Yuki Tsujikawa [1], Masafumi Horio [1], Takahiro Kondo [2,3] and Iwao Matsuda [1,*]

[1] The Institute for Solid State Physics, The University of Tokyo, Kashiwa 277-8581, Chiba, Japan; yin-heming200@g.ecc.u-tokyo.ac.jp (H.Y.)
[2] Department of Materials Science and Tsukuba Research Center for Energy Materials Science, Institute of Pure and Applied Sciences, University of Tsukuba, Tsukuba 305-8573, Ibaraki, Japan; takahiro@ims.tsukuba.ac.jp
[3] The Advanced Institute for Materials Research, Tohoku University, Sendai 980-8577, Miyagi, Japan
* Correspondence: jingmin-tang@issp.u-tokyo.ac.jp (J.T.); imatsuda@issp.u-tokyo.ac.jp (I.M.)

Abstract: Hydrogen boride (HB) sheets are emerging as a promising two-dimensional (2D) boron material, with potential applications as unique electrodes, substrates, and hydrogen storage materials. The 2D layered structure of HB was successfully synthesized using an ion-exchange method. The chemical bonding and structure of the HB sheets were investigated using Fourier Transform Infrared (FT–IR) spectroscopy and Transmission Electron Microscopy (TEM), respectively. X-ray photoelectron spectroscopy (XPS) was employed to study the chemical states and transformation of the components before and after atomic hydrogen adsorption, thereby elucidating the atomic hydrogen adsorption process on HB sheets. Our results indicate that, upon atomic hydrogen adsorption onto the HB sheets, the B-H-B bonds were broken and converted into B-H bonds. This research highlights and demonstrates the changes in chemical states and component transformations of the boron element on the HB sheets' surface before and after atomic hydrogen adsorption, thus providing a clearer understanding of the unique bonding and structural characteristics of the HB sheets.

Keywords: 2D material; hydrogen boride sheets; atomic hydrogen adsorption; photoelectron spectroscopy

1. Introduction

Owing to the unique properties and promising applications of two-dimensional materials, researches in this field have garnered significant interest and attention [1–5]. Since the successful discovery and study of various 2D materials, such as graphene (a single layer of carbon atoms [6–8]), numerous efforts have been made to produce and synthesize a diverse range of 2D materials with varying physical properties, including as insulators, semiconductors, and superconductors [2–5]. Combining these two-dimensional materials can substantially enhance the performance of electronic devices, create novel materials with intriguing new physical properties, and enable the exploration of quantum phenomena [6–12]. For instance, transistors, which are crucial components in electronic devices that control electrical signals, can be fabricated from 2D materials [8–10]. These nanoscale transistors benefit from the planar surfaces of 2D materials, which are essential for optimal performance.

As the neighboring element to carbon in the periodic table, boron, with one less electron and a lower atomic mass than carbon, can exist in different allotropes and structures, such as bulk boron, borophene, boron clusters, and even one-dimensional boron chains [13–16]. In comparison to the well-known two-dimensional carbon materials, including graphite and graphene, boron is capable of forming various 2D structures akin to those of carbon [1,13–17]. Consequently, 2D boron-based materials continue to stimulate increasing attraction and research. Additionally, metallic boride crystals, such as MgB_2, exhibit stable planar 2D structures, providing evidence for the potential synthesis of boron-based 2D materials analogous to graphene [18,19]. Borophene, a promising candidate for ideal 2D boron-based materials, also forms a layered structure with a few hexagonal boron layers. It was first synthesized on Ag (111) in 2015 [20] and has been produced on other metal surfaces [21–24]. Scanning tunneling microscopy (STM) images and theoretical calculations have confirmed the layered, planar structure of these boron materials [15,20,21]. After synthesizing and confirming the structure of borophene, studies have also explored its interlayer stacking properties [25–27]. However, an unresolved issue is that borophene reportedly cannot remain stable without a metal substrate and is reactive with water and oxygen in the air [28,29].

To avoid the unstable characteristics of borophene, hydrogen boride (HB) sheet, which is also a boron-based 2D material, has attracted more interest for its stability even in an aqueous condition [30]. The ideal structure of HB sheets has been theoretically calculated using first-principles methods to determine their theoretical discharge capacity. Results indicate that the discharge capacity of HB sheets is exceptionally high compared to graphite anodes [31–36]. Additionally, the stability of HB sheets has been examined under aqueous conditions, demonstrating that they remain stable [36]. Studies on the hydrogen storage capability of HB sheets are also advancing, with investigations into the release of gaseous H_2 under photoirradiation [37]. Furthermore, HB sheets are considered as excellent substrates for the formation of highly dispersed Ni nanoclusters [38]. Building on previous work, reducing the byproduct of boric acid could be a valuable and meaningful research avenue, potentially enhancing the purity and quality of HB sheets [39].

To understand the properties of materials, X-ray photoelectron spectroscopy (XPS) is commonly employed and often combined with complementary techniques, such as Fourier Transform Infrared (FT–IR) spectroscopy and Transmission Electron Microscopy (TEM) imagery [37–40]. By analyzing the B 1s core-level spectra of boron materials, the typical peak positions and assignments for various boron species have been reported and studied, providing a reliable reference for boron XPS research [37–41]. XPS can also be used to track changes in chemical states following different reactions, including oxidation, reduction, and adsorption processes on material surfaces. These changes are reflected in shifts in peak positions and variations in peak area ratios in the XPS spectra. Alterations in component peaks can indicate possible reactions occurring during the process and provide insights into the species changes of HB sheets and bulk boron. The purpose of this research is to focus on the chemical bonding and chemical states of the boron atoms on the HB sheets' surface, and to verify the possible reactions caused by the boron species' bonding, by applying XPS.

2. Materials and Methods

Hydrogen boride sheets are prepared via an ion-exchange reaction. Specifically, 1 g of MgB_2 black crystal powders (99%, Rare Metallic Co., Ltd., Tokyo, Japan) was stirred at 350 rpm with 45 mL of HCl-treated ion-exchange resin (Amberlite IR120B H HG, Organo Corp., Tokyo, Japan) for 3 days at room temperature (293 K). The solvent used in this process was 300 mL acetonitrile (99.5%, JIS special grade, FUJIFILM Wako Pure Chemical Corp, Osaka, Japan) in a Schlenk flask under an argon atmosphere. It is crucial to avoid exposing the reactants to air, as the process is sensitive to water, which can cause hydrolysis. After stirring for 3 days, the mixture was cooled to 273 K overnight to allow the deposition of unreacted MgB_2, ion-exchange resin, and byproduct boric acid. The mixture was then filtered using a 0.2 μm membrane filter paper under a flowing argon atmosphere. The resulting HB sheets were obtained by drying the solvent in an oil bath at 343 K, with a cold trap to remove the evaporated acetonitrile.

FT–IR spectroscopy was measured in Ar atmosphere under room temperature with the scanning range of 400–4000 cm^{-1}. The FT–IR results were recorded by the attenuated total reflectance (ATR) method using prism holders. TEM images were obtained in vacuum condition under acceleration electron-beam voltage at 200 kV.

Analysis and measurement of chemical states of the elements were performed by XPS at beamline (BL-13B) at the Photon Factory, KEK, at room temperature with photon energies of 285 and 630 eV; photon energy resolution ($E/\Delta E$) of BL-13B at the silt width of 30 μm was estimated as 10,000 at photon energy of 64.1 eV and 7200 at 867 eV [42]. Both bulk boron and HB sheets were prepared under an Ar atmosphere and baked under 373 K for 3 days to degas the remaining moisture on the surfaces. The XPS electron analyzer was equipped in the main UHV chamber (Gamma Data/Scienta, SES200) and the pass energy of the measurements was 75 eV. The reproducibility experiment was held at beamline (BL08U) at Nano Terasu.

Both bulk boron and HB sheets were exposed to the atomic hydrogen atmosphere, which can be obtained by cracking the molecular hydrogen gas by a heated tungsten filament [43], with cracking condition is under a pressure of 5.0×10^{-6} Torr and at room temperature. The heating filament current was set at 6.5 A for 10 min. For both HB sheets and bulk boron, the atomic hydrogen-adsorbed volume was set to 3000 Langmuir (L).

3. Results and Discussion

3.1. FT–IR and TEM

Figure 1a shows the HB sheets synthesized through an ion-exchange reaction involving metal borides and protons, with MgB_2 crystals as the metal boride used in this study.

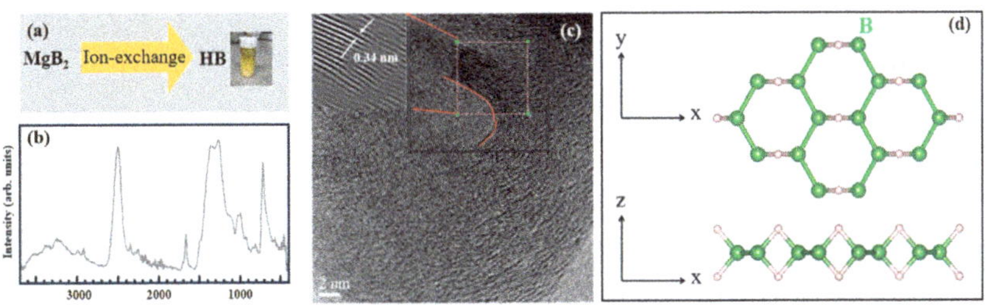

Figure 1. (**a**) Schematic synthesizing process of HB sheets by ion-exchange. (**b**) FT–IR spectrum of HB sheets. (**c**) TEM images of HB sheets; inset shows the region enclosed by the red square. A winding structure is traced by a red curve. (**d**) Structure model of HB sheets.

To examine the structure and study the properties of the HB sheets, FT–IR and TEM were employed, with the results shown in Figure 1b,c, respectively. Figure 1d displays the 3D structure model of the HB sheets [1,37–41,44]. Figure 1b presents the FT–IR spectrum of the HB sheets, which exhibits a prominent absorption peak at 2500 cm^{-1}, attributed to a stretching mode at the terminal B-H bonds at sheet edges. Peaks at 1400, and 1650 cm^{-1} correspond to bridging vibration modes and the peak at 700 cm^{-1} is attributed to the angular oscillation at the B-H-B bond, while the peak at 1000 cm^{-1} is associated with a stretching mode of the B-H bond with angular oscillations. The spectrum also shows an O-H stretching peak above 3000 cm^{-1}, indicating the presence of residual boric acid [37–41]. Figure 1c displays the TEM image, which confirms the sheet-like structure of the HB. By focusing on the enclosed area, the interlayer distance was measured to be approximately 0.34 nm, which was consistent with previous studies [38]. Additionally, the TEM image reveals a winding structure highlighted by the red curve, suggesting that the HB sheets are not as flat as the structure model. Both the FT–IR spectrum and TEM image confirm the successful synthesis of the HB sheets.

Analyzing the structure of the HB sheets from Figure 1b,c, they consist of layered structures, similar to those in an MgB_2 crystal [37–41]. Observations of vibrational modes in the FT–IR spectrum further confirm that there are B-H-B and B-H bonds within the HB sheets. The B-H-B bond can be characterized as a three center/two electron (3c–2e) bond at either surface side of the HB sheet [37–41]. As shown in Figure 1c, the HB sheets are predominantly fragmented and exhibit a winding structure.

3.2. XPS

XPS is a powerful technique for studying and characterizing the chemical elements on the surface of samples. Additionally, XPS B 1s core-level spectra can reveal changes in charge states after chemical reactions. By calculating the atomic charges of B and H atoms in HB sheets using Löwdin population analysis, the result indicated that the B atoms in the pristine HB sheets were negatively charged, as their number was slightly higher than the total valency [30]. This conclusion can also be confirmed by the observation that, if B atoms were positively charged, negatively charged O atoms of water would be expected to be closer to B atoms, rather than positively charged H atoms [30]. Based on these results, we can predict that, in the XPS B 1s core-level spectra of HB sheets, the peaks for B species (B-H-B and B-H) should appear in a lower binding energy region compared to the B 1s core-level spectra of bulk boron, which typically shows B-B bonding.

Figure 2a displays the XPS B 1s core-level spectra of the HB sheets, measured at 285 eV. Figure 2b shows the background-subtracted spectra for both the initial and 3000 Langmuir (L) atomic hydrogen-adsorbed HB sheets. The background subtraction was performed using the Shirley type function. In Figure 2b, the pristine data are presented as black lines, and the curve-fitted XPS spectra are shown as red dashed lines. Fitting of an XPS spectrum was made based on a function that combines the Gaussian (G) and Lorentzian (L) functions with software CASAXPS [45,46]. In line shape analysis, a model for fitting asymmetric peaks as a Lorentzian form convoluted with a Gaussian form is necessary, since most of the broadening in core-level spectra arises from the analyzer, X-ray source, and other final-state interactions, which are best represented by Gaussian distributions [47,48].

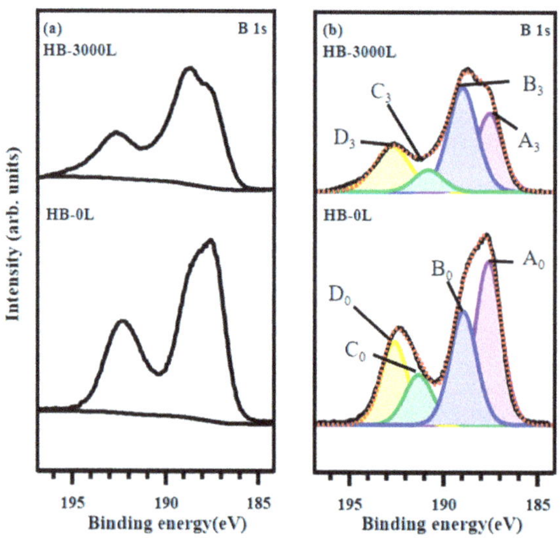

Figure 2. XPS B 1s core-level spectra for the HB sheets with photon energy of 285 eV. (**a**) Background subtraction of XPS B 1s spectra: 3000 L atomic hydrogen adsorbed on HB sheets (**above**) and initial HB sheets (**below**). (**b**) Curve fitting of the B 1s XPS spectra.

Both the initial and atomic hydrogen-adsorbed spectra are fitted with four component peaks, and the line shape of the fitting peaks is GL (30) for 30% Lorentzian and 70% Gaussian. These peaks are labeled A_0–D_0 for the initial state and A_3–D_3 for the 3000 L hydrogen-adsorbed state. The different letters indicate distinct species, while the subscripts denote the condition of atomic hydrogen adsorption, with 0 representing 0 L and 3 representing 3000 L. Peak positions, assignments, and area ratios are summarized in Table 1. Based on analysis of binding energies, taken at the measurement conditions, the four peaks can be attributed to boron species at B-H-B bonding site in HB ($A_{0,3}$), at B-H bonding site in

HB ($B_{0.3}$), at B-OH bonding site in boric acid, $B(OH)_3$ ($D_{0.3}$), and at a site of an intermediate boron material between HB and $B(OH)_3$ ($C_{0.3}$) [37–41,49–52].

Table 1. XPS fitting results for the HB sheets and bulk boron for initial and atomic hydrogen-adsorbed conditions with photon energy of 285 eV.

Samples	Component	Peak Position (eV)	Peak Assignment	Area Rate%	FWHM (eV)	Area Relative Error %
HB-0 L	A_0	187.67	B-H-B bonding site in HB	40.78	1.70	0.08
	B_0	189.02	B-H bonding site in HB	25.05	1.70	0.06
	C_0	191.47	B-OH bonding site in intermediate	7.69	1.70	0.08
	D_0	192.80	B-OH bonding site in $B(OH)_3$	26.48	1.70	0.12
HB-3000 L	A_3	187.60	B-H-B bonding site in HB	24.81	1.84	0.09
	B_3	189.10	B-H bonding site in HB	43.63	1.84	0.17
	C_3	191.00	B-OH bonding site in intermediate	7.49	1.86	0.17
	D_3	192.95	B-OH bonding site in $B(OH)_3$	24.07	1.84	0.17
Bulk boron-0 L	A_0	188.11	B-B bonding site in boron	21.82	1.87	0.16
	B_0	189.41	B-O bonding site in oxidized boron	12.47	1.89	0.17
	C_0	193.11	B-OH bonding site in oxidized boron	65.71	1.92	0.13
Bulk boron-3000 L	A_3	188.00	B-B bonding site in boron	21.01	2.14	0.17
	B_3	189.85	B-O bonding site in oxidized boron	3.36	2.12	0.17
	C_3	193.52	B-OH bonding site in oxidized boron	74.63	2.14	0.18

Further analysis of the results shown in Figure 2 reveals that both the initial and atomic hydrogen-adsorbed HB sheets exhibit a decrease in the A peak. The reduction in an area ratio of the A peak indicated a decrease in the B-H-B bridging network structure on the surface of the HB sheets. Concurrently, the increase in the B peak, assigned to the B-H bond, suggests that the breaking of the B-H-B bonds results in the formation of the B-H bonds by the hydrogeneration. The C and D peaks are located at a binding energy region of positively charged boron, considered as B-OH bonds in oxidized boron, such as boric acid. Positions of the D peak can originate from $B(OH)_3$, while those of the C peak may be from B(OH) or $B(OH)_2$ sites in the HB sheet.

To extensively explore changes in chemical state in the HB sheet, we also investigated the XPS B 1s core-level spectra with a photon energy of 630 eV. The measurement condition provides the bulk-sensitive information in XPS, as compared to that of the photon energy at 285 eV. The XPS B 1s core-level spectra of HB sheets, taken at a photon energy of 630 eV, are presented in Figure 3a,b. Figure 3a shows the background-subtracted spectra for both the pristine and 3000 L atomic hydrogen-adsorbed HB sheets, with background subtraction performed using the Shirley type. Figure 3b displays the experimental data as black lines and the curve-fitted spectra as red dashed lines. Both sets of curve-fitted spectra are also decomposed into four component peaks.

The four component peaks are labeled as A_0–D_0 for the pristine state and A_3–D_3 for the 3000 L atomic hydrogen-adsorbed state. Comparison with Figure 2b shows that the peak binding energy positions are consistent, indicating that the peak fitting is accurate, and the identified species are consistent across different photon energies. Assignments and area ratios of the four species of peaks are summarized in Table 2.

To further discuss the results shown in Figure 3, a decrease in the A peak is also observed, which indicates a reduction in the B-H-B bridging network structure on the surface of the HB sheets. Simultaneously, the increase in the B-H peak area ratio confirms that the breaking of B-H-B bonds leads to the formation of additional B-H bonds after hydrogenation. From Table 2, we observe a decrease in the A peak and increases in both the B and D peaks. Compared to the spectra obtained with 285 eV photon energy, the increase in the D peak is pronounced at 630 eV. This enhanced increase is attributed to the presence of oxygen in the bulk structure of the HB sheets, which reacts with the broken B-H-B bridging bonds, resulting in a greater formation of positively charged boron and boric acid compared to the 285 eV spectra. The results confirm that atomic hydrogen adsorption breaks the B-H-B bonds, forming more B-H bonds. Additionally, oxygen in the deeper bulk structure of the HB sheets reacts with the broken bridging B-H-B bonds, leading to a notable increase in boric acid. Moreover, the slight increase in B-H peak area ratio indicates the surface sensitive characteristics of the B-H bonding. For our experiment, since the incoming reactant is atomic hydrogen, instead of molecular hydrogen, the reaction process can be described as an Eley–Ridealtype. The Eley–Rideal mechanism is a direct reaction between the incoming molecule and the reactant at a surface, followed by the product desorption [53–55]. In the current case, atomic hydrogen twice reacts with the bridging B-H-B bonds, resulting in new B-H bonds.

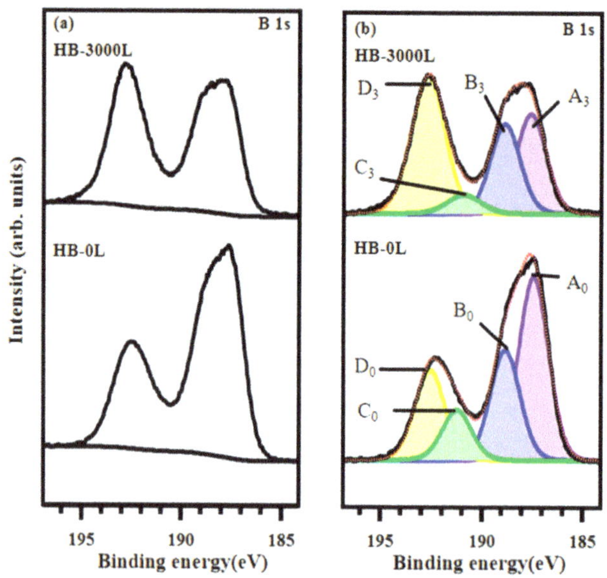

Figure 3. XPS B 1s core-level spectra for the HB sheets with photon energy of 630 eV. (**a**) Background subtraction of XPS B 1s spectra: 3000 L atomic hydrogen adsorbed on HB sheets (**above**) and initial HB sheets (**below**). (**b**) Curve fitting of the B 1s XPS spectra.

Table 2. XPS fitting results for HB sheets and bulk boron for initial and atomic hydrogen-adsorbed conditions with photon energy of 630 eV.

Samples	Component	Peak Position (eV)	Peak Assignment	Area Rate%	FWHM (eV)	Area Relative Error %
HB-0 L	A_0	187.50	B-H-B bonding site in HB	43.02	1.68	0.14
	B_0	188.87	B-H bonding site in HB	25.05	1.68	0.09
	C_0	191.21	B-OH bonding site in intermediate	7.69	1.68	0.08
	D_0	192.52	B-OH bonding site in $B(OH)_3$	26.48	1.68	0.11
HB-3000 L	A_3	187.58	B-H-B bonding site in HB	26.51	1.92	0.11
	B_3	188.74	B-H bonding site in HB	25.98	1.92	0.10
	C_3	190.91	B-OH bonding site in intermediate	5.83	1.94	0.10
	D_3	192.61	B-OH bonding site in $B(OH)_3$	41.67	1.92	0.13
Bulk boron-0 L	A_0	188.26	B-B bonding site in boron	32.77	2.27	0.15
	B_0	189.90	B-O bonding site in oxidized boron	12.14	2.27	0.17
	C_0	192.91	B-OH bonding site in oxidized boron	55.09	2.27	0.15
Bulk boron-3000 L	A_3	188.00	B-B bonding site in boron	32.57	2.16	0.30
	B_3	189.85	B-O bonding site in oxidized boron	10.75	2.16	0.28
	C_3	193.52	B-OH bonding site in oxidized boron	56.68	2.08	0.21

To verify that B-H bonds are formed by breaking of the B-H-B bonds rather than by direct interaction of boron with atomic hydrogen, we applied hydrogeneration to the bulk boron and made XPS measurements to compare the results with the HB sheets. The XPS B1s core-level spectra of bulk boron at 285 eV are shown in Figure 4a,b. The curve-fitted spectra are decomposed into three component peaks, labeled A_0–C_0 for initial state and

A_3–C_3 for 3000 L hydrogen-adsorbed state. The peak positions, the assignments, and the area ratios are given in Table 1.

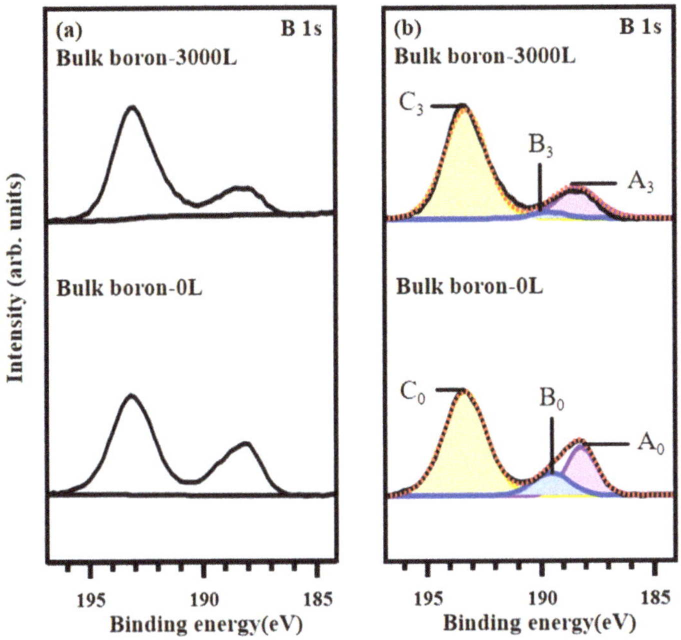

Figure 4. XPS B 1s core-level spectra for bulk boron with photon energy of 285 eV. (**a**) Background subtraction of XPS B 1s spectra: 3000 L atomic hydrogen adsorbed on HB sheets (**above**) and initial HB sheets (**below**). (**b**) Curve fitting of the B 1s XPS spectra.

In the spectra with 285 eV photon energy shown in Figure 4b, it is evident that the bulk boron sample does not react with atomic hydrogen, as the peak positions for each component (A to C) remain unchanged. This lack of change in peak positions suggests that the chemical state of boron on the surface of bulk boron does not alter after exposure to atomic hydrogen. Additionally, the area ratios of each peak show minimal variation following atomic hydrogen exposure. Specifically, the decrease in the B-O bonding peak area ratio and the increase in the boric acid peak area ratio suggest that some transformation reactions between B-O and boric acid might occur during atomic hydrogen treatment. These results indicate that the formation of new B-H bonds in the HB sheets is not a direct result of boron interacting with atomic hydrogen. Instead, it arises from the breaking of B-H-B bridging bonds, which are not present in the structure of bulk boron.

The XPS B 1s core-level spectra of bulk boron at 630 eV are also presented in Figure 5a,b. Figure 5a shows the background-subtracted spectra for both the initial and 3000 L atomic hydrogen-adsorbed bulk boron. In Figure 5b, the raw data are depicted as black lines, and the curve-fitted spectra are shown as red dashed lines. Both sets of curve-fitted spectra are decomposed into three component peaks. Analysis of the binding energies and measurement conditions reveals that the three species peaks are attributed to B-B bonds, B-O bonds, and boric acid. Compared to Figure 4b, the background interference peak is absent at 630 eV, while the positions of the other three component peaks are consistent with those observed at 285 eV. The peak positions, assignments, and area ratios are summarized in Table 2.

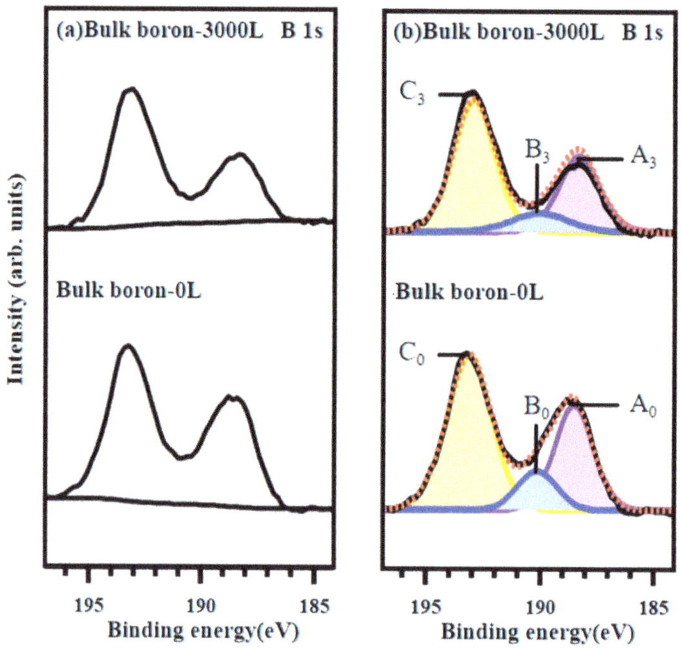

Figure 5. XPS B 1s core-level spectra for bulk boron with photon energy of 630 eV. (**a**) Background subtraction of XPS B 1s spectra: 3000 L atomic hydrogen adsorbed on HB sheets (**above**) and initial HB sheets (**below**). (**b**) Curve fitting of the B 1s XPS spectra.

The spectra obtained with 630 eV photon energy, shown in Figure 5b, also indicate that the bulk boron sample does not react with atomic hydrogen, as the peak positions for each component (A to C) remain unchanged. This lack of change in peak positions agrees with the results under 285 eV, which suggests that the chemical state of boron on the surface of the bulk boron does not alter after atomic hydrogen adsorption. Moreover, the area ratios of each peak show minimal variation following atomic hydrogen treatment. These results further confirm that the formation of new B-H bonds is not due to a direct reaction between boron and atomic hydrogen. Instead, it arises from the breaking of B-H-B bridging bonds, which are absent in the bulk boron structure.

The present method of atomic hydrogen adsorption paves a way to control a ratio between numbers of the B-H-B and B-H bonds in a HB sheet. It has been reported that the optical properties of the HB sheets depend on the B-H-B/B-H bond ratio [56]. An increase in the ratio enhances the luminescence intensity, while a decrease enhances the photoinduced H_2 release rate under UV irradiation. We have found that the hydrogenation reduces the B-H-B/B-H bond ratio and thus the method is suitable to tune the HB sheets for the optimized H_2 generation for usage. The present research has given not only academic insights into understanding the nature of the hydrogen surface chemistry at the atomic sheet but has also provided technological seeds for hydrogen production technology.

4. Conclusions

Hydrogen boride sheets were successfully synthesized via an ion-exchange reaction. Their structure and chemical bonding were examined using FT–IR and TEM. XPS B 1s core-level spectra of the HB sheets measurements were conducted at photon energies of 285 eV and 630 eV. For the surface sensitive measurement with 285 eV, the B 1s XPS spectra of the HB sheets were fitted into four component peaks, corresponding to B-H-B bonds in HB, B-H bonds in HB, B-OH bonds at the intermediate, and B-OH bonds in $B(OH)_3$, respectively.

After atomic hydrogeneration, the area ratio of the B-H-B bonding peak decreased, and the B-H bond peak increased. This result indicated that atomic hydrogen treatment led to the cleavage of the pristine B-H-B bonds and to the formation of the B-H bonds. For bulk sensitive measurement with a photon energy of 630 eV, decrease in a number of the B-H-B bond was also observed, but a number of the B-H bond increased slightly. This indicated rather that the B-H bonding formation was held at a surface. The XPS results also indicated that the breaking of B-H-B bonds led to an increase in the positively charged boron species, which suggested that the cleavage of B-H-B bonds has allowed its breakdown by the oxygen impurities in the internal bulk. In the case of bulk boron, XPS measurements revealed that the neutral boron atoms hardly reacted with the atomic hydrogen. The present study provides valuable insights into the chemical sensitivity of the B-H-B bond in HB sheets and paves a way to explore industrial reactions.

Author Contributions: Investigation, measurement, data analysis, writing original manuscript: H.Y.; Measurement, data analysis, review writing: J.T.; Measurement: K.Y.; Measurement, hydrogen resource installation: H.S.; Measurement: Y.T.; Data analysis, review writing: M.H.; Synthesis method providing, review writing: T.K.; Methodology, data analysis, review writing: I.M. All authors have read and agreed to the published version of the manuscript.

Funding: This work was supported by JSPS KAKENHI grants (Grants JP21H05012), and by JST, CREST Grant JPMJCR21O4, Japan.

Institutional Review Board Statement: Not applicable.

Informed Consent Statement: Not applicable.

Data Availability Statement: Data are contained in the article.

Conflicts of Interest: The authors declare no conflicts of interest.

References

1. Matsuda, I.; Wu, K. *2D Boron: Boraphene, Borophene, Boronene*; Springer Nature: Cham, Switzerland, 2021.
2. Olsen, T.; Andersen, E.; Okugawa, T.; Torelli, D.; Deilmann, T.; Thygesen, K.S. Discovering two-dimension topological insulators from high-throughput computations. *Phys. Rev. Mater.* **2018**, *3*, 024005. [CrossRef]
3. Chaves, A.; Azadani, J.G.; Alsalman, H.; da Costa, D.R.; Frisenda, R.; Chaves, A.J.; Song, S.H.; Kim, Y.D.; He, D.; Zhou, J.; et al. Bandgap engineering of two-dimensional semiconductor materials. *2D Mater. Appl.* **2020**, *4*, 29. [CrossRef]
4. Tan, J.; Li, S.; Liu, B.; Cheng, H. Structure, Preparation, and Applications of 2D Material-Based Metal Semiconductor Heterostructures. *Small Struct.* **2020**, *2*, 2000093. [CrossRef]
5. Qiu, D.; Gond, C.; Wang, S.; Zhang, M.; Wang, C.; Xiong, J. Recent Advances in 2D Superconductors. *Adv. Mater.* **2021**, *33*, 2006124. [CrossRef] [PubMed]
6. Jana, S.; Bandyopadhyay, A.; Datta, S.; Bhattacharya, D.; Jana, D. Emerging properties of carbon based 2D material beyond graphene. *J. Phys. Condens. Matter.* **2021**, *34*, 053001. [CrossRef]
7. Xu, L.; Wang, R.; Miao, M.; Wei, X.; Chen, Y.; Yan, H.; Lau, W.; Liu, L.; Ma, Y. Two dimensional Dirac carbon allotropes from graphene. *Nanoscale* **2014**, *21*, 1113–1118. [CrossRef]
8. Zhang, Z.; Pham, H.D.; Perepichka, D.F.; Khaliullin, R.Z. Prediction of highly stable 2D carbon allotropes based on azulenoid kekulene. *Nat. Commun.* **2024**, *15*, 1953. [CrossRef]
9. Yu, S.; Wu, X.; Wang, Y.; Guo, X.; Tong, L. 2D materials for Optical Modulation: Challenges and Opportunities. *Adv. Mater.* **2017**, *29*, 1606128. [CrossRef]
10. Lim, H.; Yoon, S.I.; Kim, G.; Jang, A.; Shin, H.S. Stacking of Two-Dimensional Materials in Lateral and Vertical Directions. *Chem. Mater.* **2014**, *26*, 4891–4903. [CrossRef]
11. Ren, S.; Tan, Q.; Zhang, J. Review on the quantum emitters in two-dimensional materials. *J. Semicond.* **2019**, *40*, 071903. [CrossRef]
12. Sobirey, L.; Luick, N.; Bohlen, M.; Biss, H.; Moritz, H.; Lompe, T. Observation of superfluidity in a strongly correlated two-dimensional Fermi gas. *Science* **2021**, *372*, 844–846. [CrossRef] [PubMed]
13. Wade, K. The structural significance of the number of skeletal bonding electron-pairs in carbonanes, the higher boranes and borane anions, and various transition-metal carbonyl cluster compounds. *J. Chem. Soc. D* **1971**, *15*, 792–793. [CrossRef]
14. Sullenger, D.B.; Phipps, K.D.; Seabaugh, P.W.; Hugens, C.R.; Sands, D.E.; Cantrell, J.S. Boron modifications produced in an induction-coupled argon plasma. *Science* **1969**, *163*, 935–937. [CrossRef] [PubMed]
15. Feng, B.; Zhang, J.; Zhong, Q.; Li, W.; Li, S.; Li, H.; Cheng, P.; Meng, S.; Chen, L.; Wu, K. Experimental realization of two-dimension boron sheets. *Nat. Chem.* **2016**, *8*, 563–568. [CrossRef]
16. Boustani, I. New quasi-planar surfaces of bare boron. *Surf. Sci.* **1997**, *370*, 355–363. [CrossRef]

17. Jemmis, E.D.; Jayasree, E.G. Analogies between Boron and Carbon. *Acc. Chem. Res.* **2003**, *36*, 816–824. [CrossRef]
18. Jones, M.E.; Marsh, R.E. The preparation and structure of magnesium boride, MgB_2. *J. Am. Chem. Soc.* **1954**, *76*, 1434–1436. [CrossRef]
19. Ma, R.; Bando, Y.; Mori, T.; Golberg, D. Direct Pyrolysis Method for Superconducting Crystalline MgB_2 Nanowires. *Chem. Mater.* **2003**, *15*, 3194–3197. [CrossRef]
20. Mannix, A.J.; Zhou, X.F.; Kiraly, B.; Wood, J.D.; Alducin, D.; Myers, B.D.; Liu, X.; Fisher, B.L.; Santiago, U.; Guest, J.R.; et al. Synthesis of borophenes: Anisotropic, two-dimension boron polymorphs. *Science* **2015**, *350*, 1513–1516. [CrossRef]
21. Qing, Z.; Jin, Z.; Peng, C.; Baojie, F.; Wenbin, L.; Shaoxiang, S.; Hui, L.; Sheng, M.; Lan, C.; Kehui, W. Metastable phases of 2D boron sheets on Ag(111). *J. Phys. Condens. Matter.* **2017**, *29*, 095002.
22. Tai, G.; Hu, T.; Zhou, Y.; Wang, X.; Kong, J.; Zeng, T.; You, Y.; Wang, Q. Synthesis of atomically think boron films on copper foils. *Angew. Chem. Int. Ed.* **2015**, *54*, 15473–15477. [CrossRef] [PubMed]
23. Wu, R.; Drozdov, I.K.; Eltinge, S.; Zahl, P.; Ismail-Beigi, S.; Bozavic, I.; Gozar, A. Large-area single-crystal sheets of borophene on Cu(111) surfaces. *Nat. Nanotechnol.* **2019**, *14*, 44–49. [CrossRef] [PubMed]
24. Kiraly, B.; Liu, X.; Wang, L.; Zhang, Z.; Mannix, A.J.; Fisher, B.L.; Takobson, B.I.; Hersam, M.C.; Guisinger, N.P. Borophene synthesis on Au(111). *ACS Nano* **2019**, *13*, 3816–3822. [CrossRef] [PubMed]
25. Gao, N.; Wu, X.; Jiang, X.; Bai, Y.; Zhao, J. Structure and stability of bilayer borophene. The roles of hexagonal hold and interlayer bonding. *FlatChem* **2018**, *7*, 48–54. [CrossRef]
26. Karmodak, N.; Jemmis, E.D. Metal templates and boron sources controlling Borophene structures: An ab initio study. *J. Phys. Chem. C* **2018**, *122*, 2268–2274. [CrossRef]
27. Ma, F.; Jiao, Y.; Gao, G.; Gu, Y.; Bilic, A.; Chen, Z.; Du, A. Graphene-like two-dimensional ionic boron with double Dirac cones at ambient condition. *Nano Lett.* **2016**, *16*, 3022–3028. [CrossRef]
28. Hou, C.; Tai, G.; Hao, J.; Sheng, L.; Liu, B.; Wu, Z. Ultrastable Crystalline Semiconducting Hydrogenated Borophene. *Angew. Chem. Int. Ed.* **2020**, *59*, 10819–10825. [CrossRef]
29. Mozvashi, S.M.; Mohebpour, M.A.; Vishkayi, S.I.; Tagani, M.B. Mechanical strength and flexibility in α-4H borophene. *Sci. Rep.* **2021**, *11*, 7547. [CrossRef]
30. Rojas, K.I.M.; Cuong, N.T.; Nishino, H.; Ishibiki, R.; Ito, S.-I.; Miyauchi, M.; Fujimoto, Y.; Tominaka, S.; Okada, S.; Hosono, H.; et al. Chemical stability of hydrogen boride nanosheets in water. *Commun. Mater.* **2021**, *81*, 81. [CrossRef]
31. Makaremi, M.; Mortazavi, B.; Singh, C.V. 2D Hydrogenated graphene-like borophene as a high capacity anode material for improved Li/Na ion batteries: A first principle study. *Mater. Today Energy* **2018**, *8*, 22–28. [CrossRef]
32. Jena, N.K.; Araujo, R.B.; Shukla, V.; Ahuja, R. Borophane as a benchmate of graphene: A potential 2D material for anode of Li and Na-ion batteries. *ACS Appl. Mater. Interfaces* **2017**, *9*, 16148–16158. [CrossRef] [PubMed]
33. Yang, Z.; Choi, D.; Kerisit, S.; Rosso, K.M.; Wang, D.; Zahng, J.; Graff, G.; Liu, J. Nanostructures and Lithium Electrochemical Reactivity of Lithium Titanites and Titanium Oxides: A Review. *J. Power Sources* **2009**, *192*, 588–598. [CrossRef]
34. Li, N.; Chen, Z.; Ren, W.; Cheng, H. Flexible graphene-based lithium ion batteries with ultrafast charge and discharge rates. *Proc. Natl. Acad. Sci. USA* **2012**, *109*, 17360–17365. [CrossRef] [PubMed]
35. Zhu, J.; Yang, D.; Yin, Z.; Yan, Q.; Zhang, H. Graphene and graphene-based materials for energy storage applications. *Small* **2014**, *10*, 3480–3498. [CrossRef]
36. Shukla, V.; Araujo, R.B.; Jena, N.K.; Ahuja, R. Borophene's tryst with stability: Exploring 2D hydrogen boride as an electrode for rechargeable batteries. *Phys. Chem. Chem. Phys.* **2018**, *20*, 22008–22016. [CrossRef]
37. Kawamura, R.; Cuong, N.T.; Fujita, T.; Ishibiki, R.; Hirabayashi, T.; Yamaguchi, A.; Matsuda, I.; Okada, S.; Miyauchi, T.K.M. Photoinduced hydrogen release from hydrogen boride sheets. *Nat. Commun.* **2019**, *10*, 4880. [CrossRef]
38. Noguchi, N.; Ito, S.; Hikichi, M.; Cho, Y.; Goto, K.; Kubo, A.; Matsuda, I.; Fujita, T.; Miyauchi, M.; Kondo, T. Highly Dispersed Ni Nanoclusters Spontaneously Formed on Hydrogen Boride Sheets. *Molecules* **2022**, *27*, 8261. [CrossRef]
39. Ito, S.; Hikichi, M.; Noguchi, N.; Yuan, M.; Kang, Z.; Fukuda, K.; Mitauchi, M.; Matsuda, I.; Kondo, T. Effective treatment of hydrogen boride sheets for long-term stabilization. *Phys. Chem. Chem. Phys.* **2023**, *25*, 15531–15538. [CrossRef]
40. Nishino, H.; Fujita, T.; Cuong, N.T.; Tominake, S.; Mitauchi, M.; Iimura, S.; Hirata, A.; Umezawa, N.; Okata, S.; Nishibori, E.; et al. Formation and characterization of hydrogen boride sheets derived from MgB_2 by cation exchange. *J. Am. Chem. Soc.* **2017**, *139*, 13761–13769. [CrossRef]
41. Tominaka, S.; Ishibiki, R.; Fujibo, A.; Kawakami, K.; Ohara, K.; Masuda, T.; Matsusda, I.; Hosono, H.; Kondo, T. Geometrical Frustration of B-H Bonds in Layered Hydrogen Borides Access by Soft Chemistry. *Chem* **2020**, *6*, 406–418. [CrossRef]
42. Toyoshima, A.; Kikuchi, T.; Tanaka, H.; Masa, K.; Amemiya, K.; Ozawa, K. Performance of PF BL-13A, a vacuum vltraviolet and soft X-ray undulator beamline for studying organic thin films adsorbed on surfaces. *J. Phys. Conf. Ser.* **2013**, *425*, 152019. [CrossRef]
43. Sutoh, A.; Okada, Y.; Ohta, S.; Kawabe, M. Cracking efficiency of hydrogen with tungsten filament in molecular beam epitaxy. *Jpn. J. Appl. Phys.* **1995**, *34*, L1379. [CrossRef]
44. Momma, K.; Izumi, F. Vesta 3 for three-dimensional visualization of crystal, volumetric and morphology data. *J. Appl. Crystallogr.* **2011**, *44*, 1272–1276. [CrossRef]

45. Fairley, N.; Fernandez, V.; Richard-Plouet, M.; Guillot-Deudon, C.; Walton, J.; Smith, E.; Flahaut, D.; Greiner, M.; Biesinger, M.; Tougaard, S.; et al. Systematic and collaborative approach to problem solving using X-ray photoelectron spectroscopy. *Appl. Surf. Sci. Adv.* **2021**, *5*, 100112. [CrossRef]
46. Available online: http://www.casaxps.com/ (accessed on 4 December 2023).
47. Nguyen, L.; Tao, F.F.; Tang, Y.; Dou, J.; Bao, X. Understanding Catalyst Surfaces during Catalysis through Near Ambient Pressure X-ray Photoelectron Spectroscopy. *Chem. Rev.* **2019**, *119*, 6822–6905. [CrossRef]
48. Schmid, M.; Steinruck, H.P.; Gottfried, J.M.A. A New Asymmetric Pseudo-Voigt Function for More Efficient Fitting of XPS Lines. *Surf. Interface Anal.* **2014**, *46*, 505–511. [CrossRef]
49. Silva-Quinones, D.; Butera, R.E.; Wang, G.T.; Teplyakov, A.V. Solution Chemistry to Control Boron-Containing Monolayers on Silicon: Reactions of Boric Acid and 4-Fluorophenylboronic Acid with H- and Cl-terminated Si(100). *Langmuir* **2021**, *37*, 7194–7202. [CrossRef]
50. Li, H.; Jing, L.; Liu, W.; Lin, J.; Tay, R.Y.; Tsang, S.H.; Teo, E.H. Scalable Production of Few-Layer Boron Sheets by Liquid-Phase Exfoliation and Their Superior Supercapacitive Performance. *ACS. Nano* **2018**, *12*, 1262–1272. [CrossRef]
51. Guo, H.; Liu, C.; Hu, H.; Zhang, S.; Ji, X.; Cao, X.; Ning, Z.; Zhu, W.; Tian, H.; Wu, Y. Neglected acidity pitfall: Boric acid-anchoring hole-selective contact for perovskite solar cells. *Natl. Sci. Rev.* **2023**, *10*, nwad057. [CrossRef]
52. Jähnichen, T.; Hojak, J.; Bläker, C.; Pasel, C.; Mauer, V.; Zittel, V.; Denecke, R.; Bathe, D.; Enke, D. Synthesis of Turbostratic Boron Nitride: Effect of Urea Decomposition. *ACS Omega* **2022**, *7*, 33375–33384. [CrossRef]
53. Zhou, L.; Jiang, B.; Alducin, M.; Guo, H. Communication: Fingerprints of reaction mechanisms in product distributions: Eley-Rideal-type reactions between D and CD_3/Cu(111). *J. Chem. Phys.* **2018**, *149*, 031101. [CrossRef] [PubMed]
54. Xiong, K.; Zhang, L.; Zhao, B.; Jiang, B. Six-dimensional quantum dynamics of an Eley-Rideal reaction between gaseous and adsorbed hydrogen atoms on Cu(111). *Faraday Discuss.* **2024**, *251*, 457–470. [CrossRef] [PubMed]
55. Zhu, L.; Hu, C.; Chen, J.; Jiang, B. Investigating the Eley-Rideal recombination of hydrogen atoms on Cu(111) via a high-dimensional neural network potential energy surface. *Phys. Chem. Chem. Phys.* **2023**, *25*, 5479–5488. [CrossRef] [PubMed]
56. Hikichi, M.; Takeshita, J.; Noguchi, N.; Ito, S.; Yasuda, Y.; Ta, L.T.; Rojas, K.I.M.; Matsuda, I.; Tominaka, S.; Morikawa, Y.; et al. Controlling Photoinduced H_2 Release from Freestanding Borophane Sheets under UV Irradiation by Tuning B–H bonds. *Adv. Mater. Interfaces* **2023**, *10*, 2300414. [CrossRef]

Disclaimer/Publisher's Note: The statements, opinions and data contained in all publications are solely those of the individual author(s) and contributor(s) and not of MDPI and/or the editor(s). MDPI and/or the editor(s) disclaim responsibility for any injury to people or property resulting from any ideas, methods, instructions or products referred to in the content.

materials

MDPI

Article

Structural, Electronic, and Magnetic Properties of Neutral Borometallic Molecular Wheel Clusters

Saira Perveen and Nevill Gonzalez Szwacki *

Faculty of Physics, University of Warsaw, Pasteura 5, PL-02093 Warsaw, Poland; sperveen@fuw.edu.pl
* Correspondence: gonz@fuw.edu.pl; Tel.: +48-22-55-32-797

Abstract: Atomic clusters exhibit properties that fall between those found for individual atoms and bulk solids. Small boron clusters exhibit planar and quasiplanar structures, which are novel materials envisioned to serve as a platform for designing nanodevices and materials with unique physical and chemical properties. Through past research advancements, experimentalists demonstrated the successful incorporation of transition metals within planar boron rings. In our study, we used first-principles calculations to examine the structure and properties of neutral boron clusters doped with transition metals, denoted as TMB_n and TMB_{2n}, where TM = Ti, Cr, Mn, Fe, Co, Nb, or Mo and $n = 8$–10. Our calculations show that the TMB_{2n} structures, which involve sandwiching metal atoms between two rings (called the drum configuration), and clusters with the single ring configuration, TMB_n, are stable. These clusters typically have relatively large HOMO-LUMO energy gaps, suggesting high kinetic stability and low chemical reactivity. Moreover, the clusters display interesting magnetic properties, determined not only by the metal atoms but also by the induced magnetism of the boron rings. These structures have potential applications in spintronics and sensing. This work also provides a basis for studying magnetism in the one-dimensional limit.

Keywords: borometallic clusters; planar boron clusters; first-principles calculations; nanostructure design

1. Introduction

Boron, a versatile element with applications spanning many fields, has a rich history that traces its origins to the isolation of simple boranes by Stock et al. [1]. Over the past two decades, a noteworthy chapter in this narrative has unfolded with the synthesis and characterization of various low-dimensional boron nanomaterials, including nanoclusters, nanowires, nanotubes, nanobelts, nanoribbons, and monolayer crystalline sheets [2–5]. Unlike traditional bulk boron crystals found in α-, β-, and γ-rhombohedral and α-tetragonal forms, these newly crafted boron-based nanomaterials have distinctive bonding patterns [6]. This departure gives rise to captivating physical and chemical properties, making them a focal point of fascination within materials science. Of particular note is the capability of boron-based nanomaterials, such as clusters, to serve as superatoms or as fundamental building blocks for crafting nanostructures designed with novel functionalities and properties. These systems are characterized by having diverse structures, including planar, quasiplanar, bowl, cage, tubular drum-like forms, multiple ring tubes, and fullerenes [7–11]. Electron deficiency of the boron atom enables its involvement in localized and delocalized bonding patterns across various geometric configurations. The neutral B_n clusters with $n < 20$ prefer a planar or quasiplanar structure [12]. On the other hand, B_n^- clusters retain planarity (or quasiplanarity) for $n \leq 40$ [13]. The perfectly planar B_8 and B_9^- molecular wheels, with a hepta- or octacoordinated central boron atom, respectively, laid the foundation of borometallic clusters [14]. In the borometallic cluster, the central B atom is replaced by a metal atom, creating a similar doubly aromatic cluster [15,16]. Experiments in cluster beams have demonstrated that transition metal (TM) atoms can be incorporated into the center of the planar boron clusters [17–19].

In a more recent study by Zhang et al. [20], the CALYPSO structure search method was used to investigate the global minimum structures of neutral and negatively charged Ta_2B_n clusters, where n ranges from 1 to 10. Their findings revealed the formation of a B ring when n equals 6. Interestingly, the research identified a neutral Ta_2B_6 cluster exhibiting a bipyramidal configuration with enhanced stability. Chang Xu et al. [21] conducted a theoretical investigation on TM-centered double-ring clusters, $M@B_{2n}$ (where M = Ti, Cr, Fe, Ni, Zn and $n = 6, 7, 8$). They explored the factors contributing to the stability of these clusters in their study. Miao Yan et al. [22] introduced the concept of fluxional bonds in various structures, including planar B_{19}^-, tubular $Ta@B_{20}^-$, and cage-like B_{39}^-. These bonds form and break continuously on these systems' extremely flat potential energy surfaces. Ren et al. [23] used the CALYPSO method combined with density functional theory (DFT) calculations on niobium-doped boron clusters. Their research revealed that the global minima for the neutral clusters correspond to half-sandwich structures at n = 10–17 and tubular-type structures at n = 18–20. Chen et al.'s [24] study offered a comprehensive overview of the geometric configurations of both pure (B_n) and metal-doped boron clusters, which exhibit structures such as planar, nanotube, bilayer, fullerene-like, and core-shell forms across a wide range of cluster sizes (n). They also explored the potential for developing boron-based nanomaterials with specific functionalities derived from metal–boron clusters. Liu et al. [25] identified ten potential clusters made up of rings formed by 12 to 15 atoms of boron–carbon doped with heavier alkaline-earth (Ae) metals (Ae = Ca, Sr, Ba). Their chemical bond analysis reveals the important role of Ae d orbitals in facilitating covalent interactions between the central Ae atom and the surrounding boron–carbon rings.

Linear magnetic chains are becoming crucial in quantum information involving entanglement and correlation [26,27]. Carbon and boron-nitride-based nanotubes are used as support surfaces to stabilize unstable isolated atomic chains [28]. Photoelectron spectroscopy and theoretical studies identified metal-centered tubular ionic structures MnB_{16}^-, CoB_{16}^-, RhB_{18}^-, and TaB_{20}^- with a magnetic moment of 2, 2, 0, and 0 μ_B, respectively [29–32]. Neutral boron-based tubular structures, including FeB_{14}, FeB_{16}, UB_{20}, NpB_{20}, and PuB_{20}, were theoretically examined and found to have a quenched magnetic moment of $2\mu_B$ for the iron-based structures and relatively high magnetic moment of 3.69, 4.92, and 6.07 μ_B, respectively, for actinide-centered tubular structures [33,34]. These research works showed that the size and stability of the tubular structures greatly depend on the encapsulated metal atoms through the covalent bonds forming between the metal ions and the boron frame (MB_n) [35]. Metal-centered tubular boron clusters have been suggested as the building blocks for one-dimensional metallo-nanotubes [35]. However, the magnetic moments of TM-centered tubular boron clusters remain unexplored, with limited discussion on their magnetic behaviors.

In this work, using DFT calculations, we study the structure and properties of molecular borometallic wheel structures derived from planar boron B_8, B_9, and B_{10} wheel clusters. For all of these clusters, we analyze their magnetic behavior using spin-polarized calculations and analyze charge density differences and the contribution of metal and boron atoms to the density of states via the projected density of states (PDOS) analysis.

2. Computational Details

In our DFT-based study, we use the plane-wave-based Quantum ESPRESSO package [36]. The generalized gradient approximation (GGA) of Perdew–Burke–Ernzerhof (PBE) [37] is employed in the calculations combined with projector-augmented wave (PAW) pseudopotentials. Boron-based rings centered with TM atoms (TMB_n) are modeled by varying the value of n from 8 to 10 for single rings and subsequently increasing it to 16, 18, and 20 by taking the same ring twice and sandwiching the metal atom between them (TMB_{2n}). The TMB_n and TMB_{2n} clusters are shown in Figure 1. A vacuum distance of 16 Å is introduced to avoid interactions between cluster replicas. A plane-wave basis set with an energy cutoff of 70 Ry is used. The convergence threshold for the self-consistent energy calculations is 10^{-12}, and the atomic positions are optimized until the forces on atoms are smaller than 10^{-4} Ry/Bohr. The convergence threshold during phonon self-consistent calculations is 10^{-14} eV. All the computations are carried out at the Γ-point of the Brillouin zone. Assessing the stability of clusters often relies on the binding energy, E_b, as an important parameter. The calculation of E_b for the TMB_n and TMB_{2n} clusters is conducted using the following equation:

$$E_b = E(\mathrm{TMB}_i) - E(\mathrm{B}_i) - E(\mathrm{TM}),$$

where $E(\mathrm{TMB}_i)$ is the total energy of the optimized boron ring with the TM atom in the center, $E(\mathrm{B}_i)$ is the total energy of the boron cluster in the ring ($i = n$) or drum ($i = 2n$) forms, and $E(\mathrm{TM})$ is the energy of the isolated TM atom. Spin-polarized calculations are carried out to study the magnetic properties of the clusters.

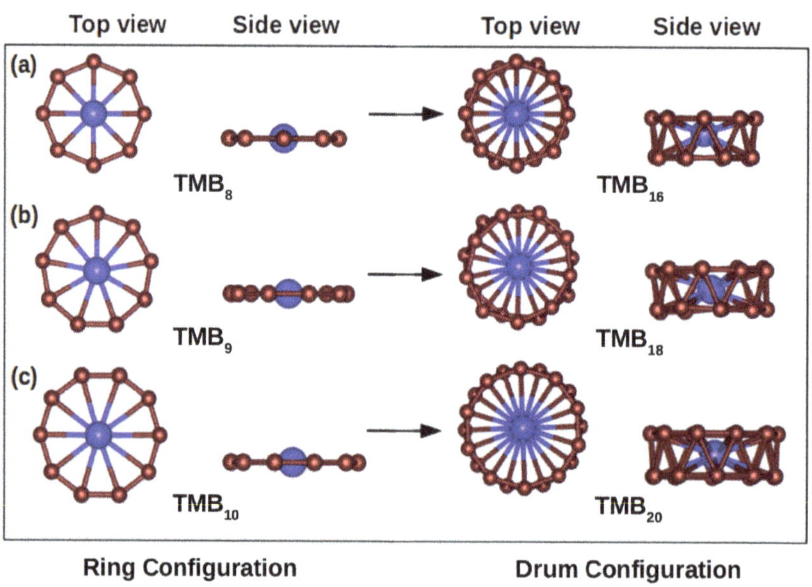

Figure 1. The structure of TMB_n (**left**) and TMB_{2n} (**right**) for $n = 8, 9$, and 10 in (**a–c**), respectively. In each case, top and side views of the clusters are shown.

3. Results and Discussion

We became interested in the high symmetry of aromatic boron B_8 and B_9^- clusters, which are perfect heptagon and octagon structures with D_{7h} and D_{8h} symmetries, respectively [38]. This led us to conduct thorough investigations involving neutral clusters formed by replacing the central B atom in the octagon boron structure with TM counterparts, specifically TM = Ti, Cr, Mn, Fe, Co, Nb, and Mo, and extending the study to larger borometallic clusters. Our investigation focused on two primary configurations: the above-mentioned "ring configuration" (TMB_n), featuring a TM atom at the center of a ring composed of 8, 9, or 10 boron atoms, and the "drum configuration" (TMB_{2n}), involving two rings of the same size stacked on top of each other, with the metal atom situated between the centers of two rings comprising 8, 9, or 10 boron atoms. Our system is visually represented in Figure 1, with boron atoms highlighted in dark red and forming a ring, and the central TM atom is shown in blue.

3.1. Structural Properties

In line with the structural arrangement shown in Figure 1, the bond lengths between adjacent boron atoms (d_{B-B}) and those connecting boron with central metal atoms (d_{B-TM}) are summarized for all the studied clusters in Table 1. It is worth noting that all the designed configurations are planar except for TiB_8, TiB_9, and FeB_8, where the central atom deviates slightly out of plane by 0.8387 Å, 0.6168 Å, and 0.8394 Å, respectively, giving rise to a distinctive bowl-like structure, which aligns well with previous research. The distances between metal and boron atoms (d_{B-TM}) for $n = 8$, 9, and 16 are small compared to their covalent bonds, indicating minimal steric repulsion between them. The bond length d_{B-B} is influenced by the number of B atoms present in the boron ring and the size of the central metal atom, especially for $n = 8$ and 9 (refer to Table 1). For TM = Cr, Mn, Fe, or Co, the drum configuration with the maximum number of boron atoms is only possible with B_{18} because, beyond that limit, an elliptical geometry prevails over a circular shape. In Table 1, we summarize the point group (PG) symmetry for each cluster, which was determined using the AFLOW software [39]. Except for the non-planar structures, the D_{4h} symmetry is assigned to TMB_8. For TMB_{10} clusters, the PG symmetry is D_{2h} due to its nearly circular geometry or C_{2h} due to its elliptical geometry. Similar trends have been reported for TaB_n^- clusters [40]: TaB_8^- forms a pyramidal structure, TaB_9^- shows increased planarity as the Ta atom integrates into a larger boron ring, and TaB_{10}^- achieves a perfectly planar, decacoordinated D_{10h} symmetry, representing the highest coordination for Ta in planar geometry, stabilized by aromaticity. Drum configurations mostly belong to the C_1 PG symmetry.

In Table 1, we have summarized the values of E_b for all the clusters. The E_b values consistently increase from TMB_8 to TMB_9 (or, in some cases, TMB_8 to TMB_{10}) and then decrease gradually until reaching TMB_{20} across all seven systems for each TM. Overall, the E_b values are relatively high for Co-doped configurations but notably lower for Mn-doped structures. It is important to note that all structures have negative E_b values, indicating the stability of the clusters since energy is required to separate them into TM and B_n.

3.2. Electronic Properties

In Table 1, we have summarized the HOMO-LUMO (H-L) energy gaps of the studied clusters. Although DFT-GGA underestimates energy gap values, most clusters still exhibit nonzero H-L energy gaps. The largest energy gaps for the ring configurations are found in CoB_9, NbB_{10}, and FeB_9, with values of 1.055 eV, 0.907 eV, and 0.785 eV, respectively. Among the drum configurations, the H-L energy gaps for MoB_{20}, CrB_{16}, and MnB_{16} (1.088 eV, 0.822 eV, and 0.804 eV, respectively) are larger compared to other drum structures. It is well established that DFT-based approaches systematically underestimate energy gaps [41]. Therefore, the true H-L energy gaps of the clusters may be even twice larger. Structures with large H-L energy gaps are associated with high kinetic stability and low chemical reactivity, making them promising candidates for future applications.

Table 1. Calculated parameters related to each studied structure: bond length between boron atoms (d_{B-B}), bond length between boron and metal atoms (d_{B-TM}), point group (PG), magnetic moment (m), H-L energy gap, and binding energy (E_b).

MB_n	d_{B-B} (Å)	d_{B-TM} (Å)	PG	m (μ_B)	H-L Gap (eV)	E_b (eV)
TiB_8	1.578	2.226	C_s	4	0.541	−9.667
TiB_9	1.558	2.335	C_s	3	0.090	−10.692
TiB_{10}	1.536	2.485	D_{2h}	2	0.617	−11.106
TiB_{16}	1.639	2.275	C_s	0	0.392	−8.214
TiB_{18}	1.608	2.460	C_1	2	0.246	−9.245
TiB_{20}	1.579	2.593	C_1	2	0.312	−7.925
CrB_8	1.587	2.073	D_{4h}	4	0.004	−11.476
CrB_9	1.551	2.269	C_s	3	0.367	−12.495
CrB_{10}	1.524	2.466	C_{2h}	2	0.357	−12.152
CrB_{16}	1.614	2.209	C_{2v}	0	0.822	−11.234
CrB_{18}	1.588	2.432	C_1	1.8	0.122	−10.471
CrB_{20}	1.608	2.645	C_1	4	0.425	−8.272
MnB_8	1.572	2.054	D_{4h}	3	0.486	−7.288
MnB_9	1.542	2.255	C_{2v}	2	0.880	−7.733
MnB_{10}	1.530	2.473	C_{2h}	3	0.264	−7.128
MnB_{16}	1.602	2.235	C_{2v}	1	0.804	−6.472
MnB_{18}	1.584	2.431	C_1	1	0.558	−5.397
MnB_{20}	1.602	2.723	C_1	3	0.439	−3.518
FeB_8	1.564	2.065	C_{4v}	2	0.016	−10.866
FeB_9	1.533	2.241	C_{2v}	1	0.785	−12.050
FeB_{10}	1.523	2.461	C_{2h}	2	0.274	−11.099
FeB_{16}	1.594	2.231	C_{4v}	1.49	0.063	−9.955
FeB_{18}	1.581	2.445	C_1	2	0.324	−8.645
FeB_{20}	1.617	2.760	C_1	2	0.268	−7.918
CoB_8	1.564	2.044	D_{4h}	1	0.001	−15.709
CoB_9	1.539	2.234	C_{2v}	0	1.055	−15.889
CoB_{10}	1.522	2.462	D_{2h}	3	0.404	−14.585
CoB_{16}	1.591	2.232	C_{4v}	1	0.297	−13.398
CoB_{18}	1.579	2.443	C_s	1	0.311	−11.828
CoB_{20}	1.619	2.749	C_1	1	0.258	−11.881
NbB_8	1.653	2.157	D_{4h}	3	0.504	−10.702
NbB_9	1.573	2.300	C_{2v}	2	0.418	−13.055
NbB_{10}	1.533	2.481	D_{2h}	1	0.907	−13.451
NbB_{16}	1.661	2.306	C_s	1	0.322	−10.318
NbB_{18}	1.561	2.477	C_1	1	0.362	−12.179
NbB_{20}	1.573	2.674	C_1	0	0.000	−10.937
MoB_8	1.628	2.126	D_{4h}	2	0.419	−12.395
MoB_9	1.560	2.281	C_{2v}	2.44	0.080	−14.021
MoB_{10}	1.524	2.467	C_s	0	0.092	−13.998
MoB_{16}	1.658	2.249	C_{2v}	0	0.616	−11.811
MoB_{18}	1.602	2.431	C_1	0	0.415	−12.663
MoB_{20}	1.568	2.760	C_1	0	1.088	−10.871

The PDOS spectra for all the clusters can be found in Section SI of the Supplementary Materials (SM). These spectra are valuable for studying the contribution of different atomic orbitals to the density of states in the studied clusters, particularly at and near the Fermi level, E_F. They also aid in visualizing the contribution of atomic orbitals to the magnetic properties of the clusters (e.g., to the shift between the spin-up and spin-down PDOS).

In general, the p-orbitals of boron and the d-orbitals of TMs make the most significant contributions to the density of states at the E_F. However, in specific cases such as TiB_9, FeB_8, CoB_8, and NbB_{20}, the contribution at the E_F comes solely from boron p-orbitals. A combination of boron p and TM d orbitals contribute to the density of states at the E_F for CrB_8, CrB_{18}, FeB_{16}, MoB_9, and MoB_{10}.

Cluster-specific observations, as inferred from the PDOS spectra, show distinct contributions to the highest occupied states, which can be summarized as follows: In the Ti series, B p-orbitals dominate in TiB_9, TiB_{10}, and TiB_{20}. In contrast, a mix of B p and TM d occurs in TiB_8, TiB_{16}, and TiB_{18}. In the Cr series, B p-orbitals significantly contribute in CrB_9 and CrB_{16}, and a mix of B p and TM d contributions occur in the rest of the Cr-based clusters. In the Mn and Fe series, the clusters frequently exhibit combined B p and TM d contributions with a consistent magnetic character. For Co-based clusters, the electronic configurations reveal that boron p-orbitals dominate the highest occupied states in CoB_8, CoB_{16}, CoB_{18}, and CoB_{20}, all of which are magnetic except for CoB_8, which lacks an H-L energy gap. In contrast, CoB_9 and CoB_{10} exhibit contributions from both boron p- and TM d-orbitals to the highest occupied states, with CoB_9 being nonmagnetic and CoB_{10} magnetic. Similarly, in Nb-based clusters, boron p-orbitals primarily contribute to the highest occupied states in NbB_8, NbB_{16}, and NbB_{18}, which are magnetic with H-L energy gaps, and NbB_{20}, which is nonmagnetic without an H-L energy gap. Clusters NbB_9 and NbB_{10} exhibit combined contributions from boron p- and TM d-orbitals, maintaining magnetic properties and distinct H-L energy gaps. Finally, in the Mo series, most of the structures are nonmagnetic (e.g., MoB_{18} and MoB_{20}), which is reflected in the symmetry of the spin-up and spin-down PDOS, contrasting with the magnetic nature of smaller clusters like MoB_9.

These observations highlight the nuanced role of TM d-orbitals and B p-orbitals in shaping the electronic and magnetic characteristics of borometallic clusters. The variability in orbital contributions underscores the tunable properties of these materials.

3.3. Vibrational Properties

The vibrational properties of the structures were analyzed using phonon computations to assess their stability. These calculations provide valuable insights into the dynamic behavior and structural integrity of the systems under study. According to the literature [42], a cluster is considered stable if all its real vibrational frequencies are positive and the minimum vibrational frequency, denoted as f_{min}, is significantly large. Table 2 presents the values of f_{min} for each cluster. For the ring configuration, TMB_8 and TMB_9 with TM = Ti, Cr, Mn, Fe, or Co have the highest values of f_{min}, whereas TMB_9 and TMB_{10} clusters have the highest f_{min} for TM = Nb and Mo. Interestingly, the FeB_{10} cluster in our calculations is at the border of stability as its f_{min} value is small but negative. For the drum configurations, there is less regularity, and large values of f_{min} are obtained for CrB_{16}, while the lowest value is for MnB_{20}. Our results for f_{min} are in close agreement with a study by Pu et al. [19], where the values of f_{min} for FeB_9 and CoB_9 were 110.5 cm^{-1} and 96.7 cm^{-1}, respectively. The complete list of phonon frequencies for each cluster is provided in Section SII of SM.

3.4. Dynamic Properties

To further verify the stability of the ring and drum structures, we performed quantum molecular dynamics (MD) simulations at two different temperatures with the choice of 1 fs for the time step. Each run was 3.5 ps long. At a temperature of 300 K, the ring and drum structures maintain their shape during the MD run, and no significant deformations occur. Both structure types deform at a temperature of 600 K but retain their wheel-like structure. This is shown in Figure 2 on the example of CoB_9 and MnB_{18}. It is worth noting that MnB_{18} has one of the smallest E_b in absolute value.

Table 2. Minimum vibrational frequencies for each studied cluster.

Ring	f_{min} (cm^{-1})	Drum	f_{min} (cm^{-1})
TiB_8	206.22	TiB_{16}	48.41
TiB_9	123.91	TiB_{18}	65.62
TiB_{10}	85.00	TiB_{20}	78.17
CrB_8	87.93	CrB_{16}	160.53
CrB_9	196.47	CrB_{18}	59.27
CrB_{10}	44.03	CrB_{20}	89.9
MnB_8	140.21	MnB_{16}	122.13
MnB_9	135.68	MnB_{18}	130.70
MnB_{10}	69.60	MnB_{20}	27.00
FeB_8	133.40	FeB_{16}	75.14
FeB_9	128.48	FeB_{18}	79.27
FeB_{10}	−22.1	FeB_{20}	64.30
CoB_8	132.22	CoB_{16}	55.24
CoB_9	102.07	CoB_{18}	70.30
CoB_{10}	78.0	CoB_{20}	90.34
NbB_8	45.69	NbB_{16}	153.38
NbB_9	93.35	NbB_{18}	38.60
NbB_{10}	87.85	NbB_{20}	102.61
MoB_8	29.26	MoB_{16}	86.96
MoB_9	95.29	MoB_{18}	41.17
MoB_{10}	68.08	MoB_{20}	106.48

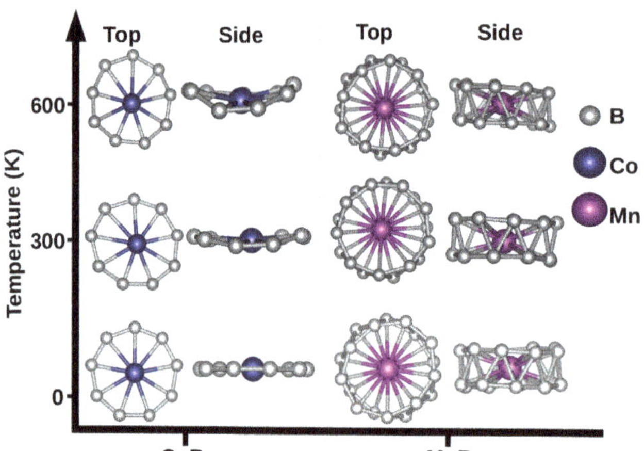

Figure 2. Ab initio molecular dynamics simulations at 300 K and 600 K. The top and side views of CoB_9 and MnB_{18} clusters are shown for each temperature.

3.5. Magnetic Properties

The magnetic moment values, denoted as *m*, for each cluster, are listed in Table 1. Generally, the value of *m* decreases as the number of boron atoms in the ring increases from 8 to 10. In contrast, an opposite trend is observed for drum structures, with *m* tending to increase as the number of boron atoms in the TMB_n structure increases (from 16 to 20). Interestingly, the *m* values for CoB_n and MoB_n drum structures are independent of *n*, remaining equal to $1\mu_B$ and $0\mu_B$, respectively.

The contribution from boron and TM atoms to magnetization is illustrated in Figure 3. In general, the contribution from TM atoms to total *m* is large. However, in most cases, the boron atoms become polarized, resulting in a non-zero contribution to m. Specifically, for TMB_8 clusters, the boron contribution (represented by the yellow line) to m is consistently larger than the TM contribution (represented by the blue line). Conversely, metals demonstrate dominance in the case of TMB_{10} structures. For drum configurations, both metal and boron atoms consistently exhibit positive (or zero) contributions to m, except for the case of Mn, where boron atoms have slightly negative local magnetization. The case of NbB_n clusters is particularly interesting, as TM atoms induce the magnetization, but m is entirely localized on the boron atoms.

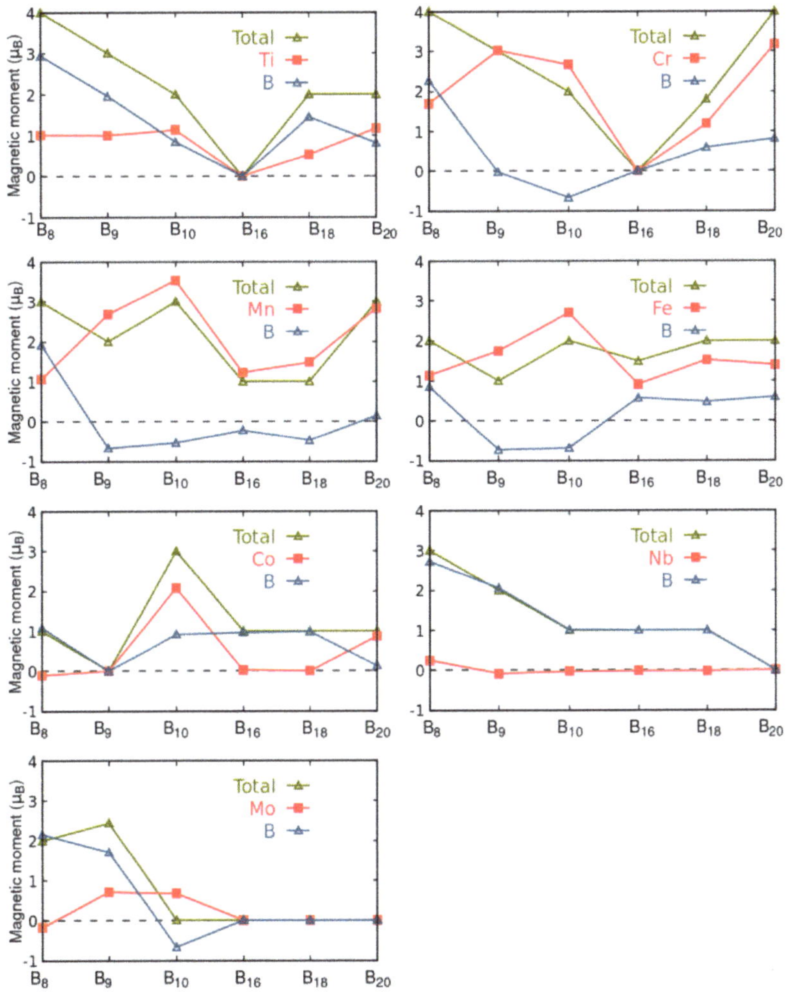

Figure 3. Total magnetic moment m and local magnetic moment contributions on the y-axis versus the cluster type in the x-axis. The contributions come from the TM (TM = Ti, Cr, Mn, Fe, Co, Nb, or Mn) atom and the boron atoms of each TMB_n cluster. In the graph, the total magnetization is represented by yellow triangles, the magnetization due to boron atoms is represented by blue triangles, and the red squares represent the contribution from the TMs. The lines are guides for the eye.

The result presented in Figure 3 can be spatially visualized by drawing the difference between the spin-up charge density ($\rho^{\uparrow}$) and the spin-down charge density ($\rho^{\downarrow}$):

$$\textit{spin polarization density} = \rho^{\uparrow}(\vec{r}) - \rho^{\downarrow}(\vec{r}).$$

This can be seen in Figure 4 where the positive and negative values of the spin polarization density are represented with yellow and cyan, respectively. This figure clearly shows that the boron atoms in all clusters become polarized, and this polarization significantly influences the total value of m. Our Löwdin population analysis provided in Section SIII of SM quantitatively demonstrates this. This analysis evidenced, in most cases, a charge transfer from/to the TM atom, which gives rise to polarization effects on the boron atoms that, as a result, contribute to the total m. The results suggest that the polarization of boron atoms is important in modulating the electron distribution, which may affect the cluster's reactivity and interactions in different environments.

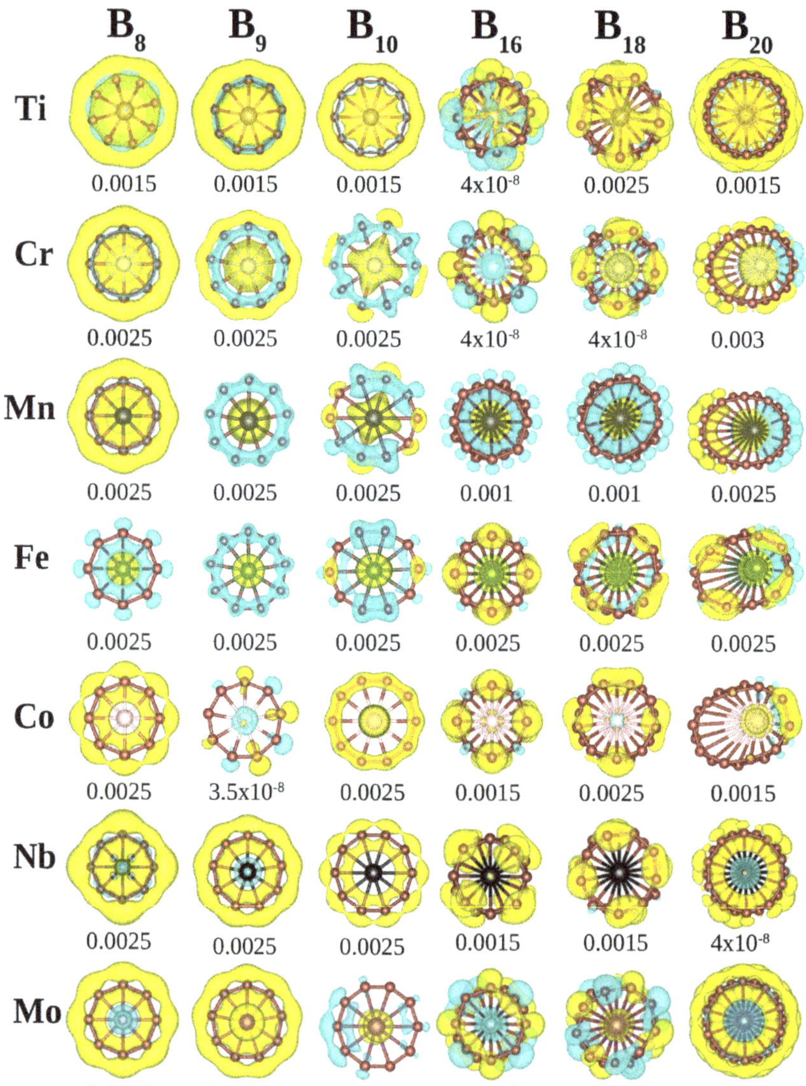

Figure 4. Difference between the spin-up charge density and the spin-down charge density. Positive and negative values are represented with yellow and cyan, respectively. The value of the iso-surface is given below each system.

4. Summary and Conclusions

The successful integration of TM atoms into the core of planar boron clusters B_n creates a fascinating class of borometallic molecules. In our study, we present the results of first-principles calculations aimed at examining the structural, electronic, and magnetic properties of boron rings with a TM atom in the center, TMB_n (with TM being Ti, Cr, Mn, Fe, Co, Nb, or Mo, and n being 8, 9, or 10), as well as boron double-rings sandwiching the metal atom, TMB_{2n}. In these molecular structures, each B atom in the circumference provides two electrons to the B–B peripheral covalent bonds and one electron to the delocalized B–TM bonds. Consequently, a TM atom with the appropriate number of valence electrons can be accommodated in the center of the boron wheel to form the TMB_n clusters, resulting in a structurally and dynamically stable borometallic cluster compound. Most of the studied clusters also possess relatively high H-L energy gaps, which may suggest low chemical reactivity. Our investigation also indicates that the overall magnetic moment of the clusters is a combination of the magnetic moment of the central TM atom and the induced magnetic moment of the peripheral boron atoms. These molecular-based magnets provide a platform for studying magnetism in the zero-dimensional limit. Furthermore, the TMB_{2n} clusters can be building blocks for one-dimensional tubular forms.

Supplementary Materials: The following supporting information can be downloaded at: https://www.mdpi.com/article/10.3390/ma18020459/s1.

Author Contributions: Conceptualization, S.P. and N.G.S.; methodology, S.P.; validation, S.P.; investigation, S.P. and N.G.S.; writing and original draft preparation, S.P.; writing, review, and editing, N.G.S.; visualization, S.P.; supervision, N.G.S. All authors have read and agreed to the published version of the manuscript.

Funding: This research received no external funding.

Institutional Review Board Statement: Not applicable.

Informed Consent Statement: Not applicable.

Data Availability Statement: The original contributions presented in this study are included in the article. Further inquiries can be directed to the corresponding author.

Acknowledgments: The use of supercomputers at the Interdisciplinary Centre for Mathematical and Computational Modelling (ICM) at the University of Warsaw is gratefully acknowledged.

Conflicts of Interest: The authors declare no conflicts of interest.

References

1. Stone, F.G.A. Chemical reactivity of the boron hydrides and related compounds. In *Advances in Inorganic Chemistry and Radiochemistry*; Elsevier: Amsterdam, The Netherlands, 1960; Volume 2, pp. 279–313.
2. Feng, B.; Zhang, J.; Zhong, Q.; Li, W.; Li, S.; Li, H.; Cheng, P.; Meng, S.; Chen, L.; Wu, K. Experimental realization of two-dimensional boron sheets. *Nat. Chem.* **2016**, *8*, 563–568. [CrossRef] [PubMed]
3. Tarkowski, T.; Gonzalez Szwacki, N.; Marchwiany, M. Structure of porous two-dimensional boron crystals. *Phys. Rev. B* **2021**, *104*, 195423. [CrossRef]
4. Tarkowski, T.; Gonzalez Szwacki, N. Boron nanotube structure explored by evolutionary computations. *Crystals* **2022**, *13*, 19. [CrossRef]
5. Tarkowski, T.; Gonzalez Szwacki, N. The structure of thin boron nanowires predicted using evolutionary computations. *Solid State Sci.* **2023**, *142*, 107241. [CrossRef]
6. Li, W.; Wu, K.; Chen, L. Epitaxial growth of borophene on substrates. *Prog. Surf. Sci.* **2023**, *98*, 100704. [CrossRef]
7. Van Duong, L.; Mai, D.T.T.; Pham-Ho, M.P.; Nguyen, M.T. A theoretical approach to the role of different types of electrons in planar elongated boron clusters. *Phys. Chem. Chem. Phys.* **2019**, *21*, 13030–13039. [CrossRef] [PubMed]
8. Pham, H.T.; Muya, J.T.; Buendia, F.; Ceulemans, A.; Nguyen, M.T. Formation of the quasiplanar B 50 boron cluster: Topological path from B 10 and disk aromaticity. *Phys. Chem. Chem. Phys.* **2019**, *21*, 7039–7044. [CrossRef] [PubMed]

9. Dong, X.; Jalife, S.; Vásquez-Espinal, A.; Ravell, E.; Pan, S.; Cabellos, J.L.; Liang, W.y.; Cui, Z.h.; Merino, G. Li2B12 and Li3B12: Prediction of the Smallest Tubular and Cage-like Boron Structures. *Angew. Chem. Int. Ed.* **2018**, *57*, 4627–4631. [CrossRef] [PubMed]
10. Van Duong, L.; Pham, H.T.; Tam, N.M.; Nguyen, M.T. A particle on a hollow cylinder: The triple ring tubular cluster B 27+. *Phys. Chem. Chem. Phys.* **2014**, *16*, 19470–19478. [CrossRef]
11. Chen, Q.; Zhang, S.; Bai, H.; Tian, W.; Gao, T.; Li, H.; Miao, C.; Mu, Y.; Lu, H.; Zhai, H.; et al. Cage-Like B41+ and B422+: New Chiral Members of the Borospherene Family. *Angew. Chem. Int. Ed.* **2015**, *54*, 8160–8164. [CrossRef]
12. Tai, T.B.; Tam, N.M.; Nguyen, M.T. Structure of boron clusters revisited, Bn with n = 14–20. *Chem. Phys. Lett.* **2012**, *530*, 71–76. [CrossRef]
13. Li, W.L.; Chen, X.; Jian, T.; Chen, T.T.; Li, J.; Wang, L.S. From planar boron clusters to borophenes and metalloborophenes. *Nat. Rev. Chem.* **2017**, *1*, 0071. [CrossRef]
14. Zhai, H.J.; Alexandrova, A.N.; Birch, K.A.; Boldyrev, A.I.; Wang, L.S. Hepta-and octacoordinate boron in molecular wheels of eight-and nine-atom boron clusters: Observation and confirmation. *Angew. Chem. Int. Ed.* **2003**, *42*, 6004–6008. [CrossRef] [PubMed]
15. Romanescu, C.; Galeev, T.R.; Li, W.L.; Boldyrev, A.I.; Wang, L.S. Aromatic metal-centered monocyclic boron rings: Co©B8- and Ru©B9-. *Angew. Chem. Int. Ed.* **2011**, *50*, 9334–9337. [CrossRef]
16. Romanescu, C.; Galeev, T.R.; Li, W.L.; Boldyrev, A.I.; Wang, L.S. Geometric and electronic factors in the rational design of transition-metal-centered boron molecular wheels. *J. Chem. Phys.* **2013**, *138*, 134315. [CrossRef]
17. Romanescu, C.; Galeev, T.R.; Li, W.L.; Boldyrev, A.I.; Wang, L.S. Transition-metal-centered monocyclic boron wheel clusters (M©B n): A new class of aromatic borometallic compounds. *Acc. Chem. Res.* **2013**, *46*, 350–358. [CrossRef] [PubMed]
18. Miao, C.; Guo, J.; Li, S. M@ B9 and M@ B10 molecular wheels containing planar nona-and deca-coordinate heavy group 11, 12, and 13 metals (M = Ag, Au, Cd, Hg, In, Tl). *Sci. China Ser. B Chem.* **2009**, *52*, 900–904. [CrossRef]
19. Pu, Z.; Ito, K.; Schleyer, P.v.R.; Li, Q.S. Planar hepta-, octa-, nona-, and decacoordinate first row d-block metals enclosed by boron rings. *Inorg. Chem.* **2009**, *48*, 10679–10686. [CrossRef] [PubMed]
20. Zhang, X.; Hu, Y.; Yuan, Y.; Li, Q.; Jiang, H.; Yang, J.; Lin, W.; Huang, H. Structure and electronic properties of neutral and anionic boron clusters doped with two tantalum atoms. *Mol. Phys.* **2022**, *120*, e2029964. [CrossRef]
21. Xu, C.; Cheng, L.; Yang, J. Double aromaticity in transition metal centered double-ring boron clusters M@B_{2n} (M = Ti, Cr, Fe, Ni, Zn; n = 6, 7, 8). *J. Chem. Phys.* **2014**, *141*, 124301. [CrossRef]
22. Yan, M.; Li, H.; Zhao, X.; Lu, X.; Mu, Y.; Lu, H.; Li, S. Fluxional Bonds in Planar B_{19}^{-}, Tubular Ta@B_{20}^{-}, and Cage-Like B_{39}^{-}. *J. Comput. Chem.* **2018**, *40*, 966–970. [CrossRef]
23. Ren, M.; Jin, S.; Wei, D.; Jin, Y.; Tian, Y.; Lu, C.; Gutsev, G.L. NbB_{12}^{-}: A new member of half-sandwich type doped boron clusters with high stability. *Phys. Chem. Chem. Phys.* **2019**, *21*, 21746–21752. [CrossRef]
24. Chen, B.; He, K.; Dai, W.; Gutsev, G.L.; Lu, C. Geometric and electronic diversity of metal doped boron clusters. *J. Phys. Condens. Matter* **2023**, *35*, 183002. [CrossRef] [PubMed]
25. Liu, X.b.; Tiznado, W.; Cui, L.J.; Barroso, J.; Leyva-Parra, L.; Miao, L.h.; Zhang, H.y.; Pan, S.; Merino, G.; Cui, Z.h. Exploring the Use of "Honorary Transition Metals" To Push the Boundaries of Planar Hypercoordinate Alkaline-Earth Metals. *J. Am. Chem. Soc.* **2024**, *146*, 16689–16697. [CrossRef]
26. Choi, D.J.; Lorente, N.; Wiebe, J.; Von Bergmann, K.; Otte, A.F.; Heinrich, A.J. Colloquium: Atomic spin chains on surfaces. *Rev. Mod. Phys.* **2019**, *91*, 041001. [CrossRef]
27. Wang, J.H.; Li, Z.Y.; Yamashita, M.; Bu, X.H. Recent progress on cyano-bridged transition-metal-based single-molecule magnets and single-chain magnets. *Coord. Chem. Rev.* **2021**, *428*, 213617. [CrossRef]
28. Ferstl, P.; Hammer, L.; Sobel, C.; Gubo, M.; Heinz, K.; Schneider, M.A.; Mittendorfer, F.; Redinger, J. Self-organized growth, structure, and magnetism of monatomic transition-metal oxide chains. *Phys. Rev. Lett.* **2016**, *117*, 046101. [CrossRef] [PubMed]
29. Jian, T.; Li, W.L.; Popov, I.A.; Lopez, G.V.; Chen, X.; Boldyrev, A.I.; Li, J.; Wang, L.S. Manganese-centered tubular boron cluster–MnB16-: A new class of transition-metal molecules. *J. Chem. Phys.* **2016**, *144*, 154310. [CrossRef] [PubMed]
30. Popov, I.A.; Jian, T.; Lopez, G.V.; Boldyrev, A.I.; Wang, L.S. Cobalt-centred boron molecular drums with the highest coordination number in the CoB16- cluster. *Nat. Commun.* **2015**, *6*, 8654. [CrossRef] [PubMed]
31. Jian, T.; Li, W.L.; Chen, X.; Chen, T.T.; Lopez, G.V.; Li, J.; Wang, L.S. Competition between drum and quasiplanar structures in RhB 18-: motifs for metallo-boronanotubes and metallo-borophenes. *Chem. Sci.* **2016**, *7*, 7020–7027. [CrossRef]
32. Li, W.L.; Jian, T.; Chen, X.; Li, H.R.; Chen, T.T.; Luo, X.M.; Li, S.D.; Li, J.; Wang, L.S. Observation of a metal-centered B 2-Ta@ B 18- tubular molecular rotor and a perfect Ta@ B 20- boron drum with the record coordination number of twenty. *Chem. Commun.* **2017**, *53*, 1587–1590. [CrossRef] [PubMed]
33. MinháTam, N.; TanáPham, H.; VanáDuong, L.; PhuongáPham-Ho, M.; ThoáNguyen, M. Fullerene-like boron clusters stabilized by an endohedrally doped iron atom: B n Fe with n= 14, 16, 18 and 20. *Phys. Chem. Chem. Phys.* **2015**, *17*, 3000–3003.

34. Wang, J.; Zhang, N.X.; Wang, C.Z.; Wu, Q.Y.; Lan, J.H.; Chai, Z.F.; Nie, C.M.; Shi, W.Q. Theoretical probing of twenty-coordinate actinide-centered boron molecular drums. *Phys. Chem. Chem. Phys.* **2021**, *23*, 26967–26973. [CrossRef]
35. Fan, Y.W.; Zhang, W.; Ge, N.N.; Li, Z. Design of Lanthanide Single-Chain Magnets Based on Tubular Segment Clusters. *J. Phys. Chem. C* **2022**, *127*, 621–626. [CrossRef]
36. Giannozzi, P.; Baroni, S.; Bonini, N.; Calandra, M.; Car, R.; Cavazzoni, C.; Ceresoli, D.; Chiarotti, G.L.; Cococcioni, M.; Dabo, I.; et al. QUANTUM ESPRESSO: A modular and open-source software project for quantum simulations of materials. *J. Phys. Condens. Matter* **2009**, *21*, 395502. [CrossRef]
37. Perdew, J.P.; Burke, K.; Ernzerhof, M. Generalized gradient approximation made simple. *Phys. Rev. Lett.* **1996**, *77*, 3865. [CrossRef]
38. Zubarev, D.Y.; Boldyrev, A.I. Comprehensive analysis of chemical bonding in boron clusters. *J. Comput. Chem.* **2006**, *28*, 251–268. [CrossRef]
39. Curtarolo, S.; Setyawan, W.; Hart, G.L.; Jahnatek, M.; Chepulskii, R.V.; Taylor, R.H.; Wang, S.; Xue, J.; Yang, K.; Levy, O.; et al. AFLOW: An automatic framework for high-throughput materials discovery. *Comput. Mater. Sci.* **2012**, *58*, 218–226. [CrossRef]
40. Li, W.L.; Ivanov, A.S.; Federič, J.; Romanescu, C.; Černušák, I.; Boldyrev, A.I.; Wang, L.S. On the way to the highest coordination number in the planar metal-centred aromatic Ta©B10- cluster: Evolution of the structures of TaB(n)- (n = 3–8). *J. Chem. Phys.* **2013**, *139*, 104312. [CrossRef]
41. Bystrom, K.; Falletta, S.; Kozinsky, B. Training Machine-Learned Density Functionals on Band Gaps. *J. Chem. Theory Comput.* **2024**, *20*, 7516–7532. [CrossRef] [PubMed]
42. Hoffmann, R.; Schleyer, P.v.R.; Schaefer, H.F. Predicting Molecules—More Realism, Please! *Angew. Chem. Int. Ed.* **2008**, *47*, 7164–7167. [CrossRef] [PubMed]

Disclaimer/Publisher's Note: The statements, opinions and data contained in all publications are solely those of the individual author(s) and contributor(s) and not of MDPI and/or the editor(s). MDPI and/or the editor(s) disclaim responsibility for any injury to people or property resulting from any ideas, methods, instructions or products referred to in the content.

materials

MDPI

Communication

Synthesis and Thermal Oxidation Resistance of Boron-Rich Boron–Carbide Material

Seth Iwan [1], Wesley Sutton [2], Paul A. Baker [1], Raimundas Sereika [1] and Yogesh K. Vohra [1,*]

[1] Department of Physics, University of Alabama at Birmingham, Birmingham, AL 35294, USA; iwanseth@uab.edu (S.I.); pabaker@uab.edu (P.A.B.); rsereika@uab.edu (R.S.)
[2] Department of Chemistry, Physics, and Astronomy, Georgia College and State University, Milledgeville, GA 31061, USA; wesley.sutton@bobcats.gcsu.edu
* Correspondence: ykvohra@uab.edu

Abstract: A boron-rich boron–carbide material ($B_{4+\delta}C$) was synthesized by spark plasma sintering of a ball-milled mixture of high-purity boron powder and graphitic carbon at a pressure of 7 MPa and a temperature of 1930 °C. This high-pressure, high-temperature synthesized material was recovered and characterized by X-ray diffraction, X-ray photoelectron spectroscopy, Raman spectroscopy, Vickers hardness measurements, and thermal oxidation studies. The X-ray diffraction studies revealed a single-phase rhombohedral structure (space group R-3m) with lattice parameters in hexagonal representation as a = 5.609 ± 0.007 Å and c = 12.082 ± 0.02 Å. The experimental lattice parameters result in a value of δ = 0.55, or the composition of the synthesized compound as $B_{4.55}C$. The high-resolution scans of boron binding energy reveal the existence of a B-C bond at 188.5 eV. Raman spectroscopy reveals the existence of a 386 cm^{-1} vibrational mode representative of C-B-B linear chain formation due to excess boron in the lattice. The measured Vickers microhardness at a load of 200 gf shows a high hardness value of 33.8 ± 2.3 GPa. Thermal gravimetric studies on $B_{4.55}C$ were conducted at a temperature of 1300 °C in a compressed dry air environment, and its behavior is compared to other high-temperature ceramic materials such as high-entropy transition metal boride. The high neutron absorption cross section, high melting point, high mechanical strength, and thermal oxidation resistance make this material ideal for applications in extreme environments.

Keywords: boron–carbon system; thermal oxidation; extreme environments; mechanical properties; Raman spectroscopy

1. Introduction

The boron-rich boron–carbide materials are of interest because of their high strength and abrasion resistance applications, their high neutron absorption cross section in nuclear reactor control applications, and their incorporation in the fabrication of light-weight composites in body armor applications. One contributor to boron's complexity is its propensity to waver between metallic or insulating properties, which arises due to frustration caused by the localization of its three valence electrons [1]. As of 2017, even boron's most understood allotropes frequently have disagreeing theoretical predictions and experimental observations of their electronic properties [2]. One industry in which boron has a wide impact is the nuclear industry. ^{10}B has a neutron cross section of 3840 barns, making it highly effective at capturing neutrons from fission chain reactions [3]. Once absorbed, the ^{10}B nucleus decays into an alpha particle and a lithium nucleus [4], which are unable to continue the fission reaction. This makes $^{10}B_4C$ an excellent option for control rods used to adjust the rate of fission in a reactor, as well as neutron shielding in general. Boron-rich boron–carbon compounds could offer high-strength materials that, due to being mostly boron, still offer high neutron absorption rates.

Boron icosahedra (B_{12}), a unique structure that forms in boron isotopes and compounds, has been studied extensively under ambient and high pressures [5,6]. In the boron–carbon system, $B_{50}C_2$ (3.8 atomic percent carbon) has been studied in detail and synthesized in a metastable state by a variety of methods [7–9]. For slightly higher atomic concentrations, approximately 8.8 atomic percent of carbon, it has been shown that boron icosahedra interconnects with tri-atom chains to form stable structures [10,11]. We are interested in the composition range between 8.8 atomic percent and 20 atomic percent carbon, where the rhombohedral $B_{4+\delta}C$ phase is expected to be stable and whose thermal and mechanical properties can be investigated in detail.

This study establishes a method of direct synthesis of boron-rich boron–carbide material ($B_{4+\delta}C$) starting from precursor material containing high-purity boron powder mixed with graphite powder at high pressures and high temperatures. A single-phase rhombohedral structure has been synthesized in bulk and characterized by a variety of analytical techniques. The thermal oxidization in air has been characterized up to 1300 °C and compared with other high-temperature ceramic materials.

2. Materials and Methods

In the synthesis process, boron powder (99.99% purity from SkySpring Nanomaterials, Houston, TX, USA) and graphite powder (99.9995% purity from Thermo Scientific, Waltham, MA, USA) were mixed to have a 3.8% atomic concentration for the intended composition of the $B_{50}C_2$ compound. For the 3.8 atomic percent precursor, 7.0005 g of boron powder was mixed with 0.3112 g of carbon powder. The powders were loaded into a tungsten carbide canister with two 7-g tungsten carbide balls and 15 mL of acetone. The container was then loaded into a SPEX, Metuchen, NJ, USA, 8000M MIXER/MILL. The material was ball-milled for 6 h in 1-h segments, with 12 min of cool-down time between each to ensure the mill did not overheat. Once finished, the canister and material were placed open into a Thermo Scientific, Waltham, USA, lab-line vacuum oven at ambient temperature for 45 min to allow the residual acetone to evaporate. The material was then removed from the canister and put through a #400 sieve to ensure no large granules were left.

Spark plasma sintering (SPS) was used to sinter and synthesize the precursor powder once it was made. SPS was performed in a DR. FRITSCH, Fellbach, Germany, FAST/SPS-Sintering Press DSP 507. A five-part die system was used, consisting of an outer cylindrical mold and an interior cylindrical mold in two pieces, as well as a top and bottom punch. The outer cylindrical mold had partially drilled holes in the sides for later use with the SPS system's pyrometer. The sample is wrapped in a molybdenum sheet and then covered with graphite foil. The molybdenum sheet was used to prevent contaminating the sample with graphite, and the graphite sheet was used to ease the removal of the molybdenum and sample from the mold after sintering was completed.

In each sample, either one or one and a half grams of material were loaded into the mold. The outside of the mold was surrounded by a graphite blanket to help thermally insulate it, with a hole cut in the blanket to allow the pyrometer to be aligned with the drilled hole in the side of the outer mold piece. A piece of graphite blanket was also used on the top of the mold, with a hole cut out for the upper punch. To protect the pistons in the SPS system, silicon–carbide plates were used on the top and bottom of the mold.

Once loaded, the sample was compressed to 3 MPa to confirm the mold would compress properly. The system was then sealed and evacuated, and nitrogen gas (95% N_2, 5% H_2) was pumped into the system at a rate of 20 L per minute. The sample was then compressed to 7 MPa and rapidly heated to its max temperature, between 1500 and 1930 °C. The average heating rate was 45 °C/min. The sample is then left for 10–20 min at max temperature before being cooled to 300 °C. The average cooling rate was 50 °C/min. Once cooled to 300 °C, the nitrogen gas was cut off and the chamber was purged. The sample was then removed from the chamber to finish cooling.

Polishing was used to clean the samples post-sintering. Once the sample was fully cooled and removed from the mold, the outer layers of molybdenum and graphite had to be removed. The graphite layer could be removed with a straight edge, while significant portions of the molybdenum sheet would have to be removed from the sample by polishing. The samples were polished using diamond grit sandpaper at 300 rpm. This was used to prepare the sample for XRD by flattening the top and bottom surfaces of the sample parallel to each other and making sure the sides were perpendicular to the top and bottom surfaces. Several samples were also further polished for microhardness testing. This was carried out by casting the samples in quick-cure epoxy with one surface of the sample exposed. The surface was then polished with increasing grit sandpaper to produce a smooth, flat surface. The grits used were 180, 320, 400, 600, 800, 1200, and 1500.

X-ray diffraction (XRD) was performed on the samples once polished. The samples were studied using powder XRD in an Empyrean X-ray Diffraction System, which was manufactured by Malvern Panalytical B.V., Almelo, Neverlands. The resulting XRD data was analyzed using High-Score plus (version 4.8) and GSAS II. The XRD patterns were collected on boron and graphite precursors as well as synthesized $B_{4+\delta}C$ samples.

Thermal oxidization measurements were conducted during a thermogravimetric analysis (TGA) stage. The Linsies Robbinsville, USA, STA PT 1600(TGA-DSC) is capable of reaching a temperature of 1600 °C while controlling the atmospheric conditions inside the furnace. A type-k thermocouple was used to read the sample temperature inside the Al2O3 crucible. The type-k thermocouple is a long, thin alumina rod that lifts the crucible above the high-precision beam balance and into the furnace. As temperature changes, thermal expansion will cause the long stem to bend, resulting in a change in the center of mass, which is seen as a perceived mass change. This bending is repeatable and is accounted for by creating zero curves, which are made by cycling the TGA stage through the desired heating profile with the empty crucible repeated 3–5 times. These curves are then averaged together and subtracted from the actual measurement curve.

3. Results

Figure 1 shows the $B_{4+\delta}C$ sample synthesized at 7 MPa and 1930 °C and recovered from the sample assembly in the spark plasma sintering system.

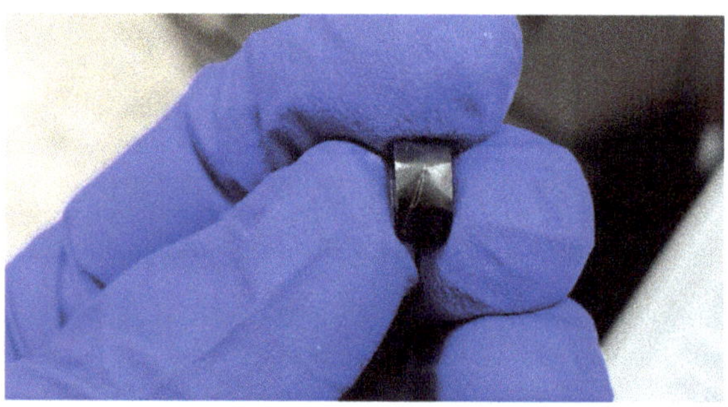

Figure 1. Handheld $B_{4+\delta}C$ sample synthesized by the spark plasma sintering process. The image shown is during the polishing process, and the silver-colored streaks are from molybdenum foil surrounding the sample that still needs to be removed.

Figure 2 shows the measured X-ray powder diffraction pattern of the synthesized sample with an X-ray wavelength of λ = 1.5406 Å. Figure 2 also shows Rietveld refinement, resulting in a fit to the rhombohedral structure (space group R-3m) with lattice parameters in hexagonal representation as a = 5.609 ± 0.007 Å and c = 12.082 ± 0.02 Å (PDF#35-0798 from the JCPDS data base). There is also a small amount of MoB_2 phase (approximately 3%) present in the Rietveld refinement resulting from the formation of molybdenum diboride between the sample powder and molybdenum foil during the spark plasma sintering process. The residual MoB_2 is on the surface of the sample and can be removed with further polishing. The fitted lattice parameters of the MoB_2 impurity phase are a = 3.038 Å and c = 3.068 Å.

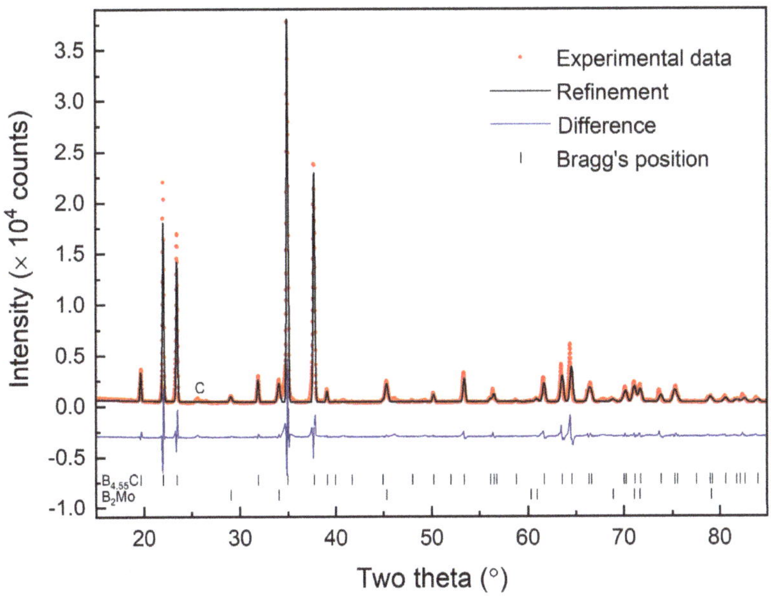

Figure 2. The measured X-ray diffraction pattern for the synthesized $B_{4+\delta}C$ sample recorded with an X-ray wavelength of λ = 1.5406 Å. Rietveld refinement of the X-ray diffraction results in the assignment of a rhombohedral phase of $B_{4+\delta}C$ with a lattice parameter described in the text. A very weak (3%) MoB_2 phase was detected due to the small amount of impurity from the sample container in the synthesis process. A very low-intensity diffraction peak from free carbon in the sample is labeled as (C).

The lattice parameters of the rhombohedral phase of $B_{4+\delta}C$ compounds have been extensively studied in the range of carbon from 8.8 atomic percent to 20 atomic percent [12]. In Figure 3, we show our measured lattice parameters and unit cell superimposed on the data from Ref. [12]. From this superposition, we estimate our atomic concentration of carbon to be 18%, or the composition of the compound to be $B_{4.55}C$. We started with 3.8 atomic percent carbon in the precursor material; however, molybdenum foil surrounding the sample serves as a sink of boron, and a carbon rich $B_{4.55}C$ phase is formed during the spark plasma sintering process. Such a determination is consistent with the Raman spectroscopy data described later in this paper.

Figure 4 shows X-ray photoelectron spectroscopy (XPS) of the $B_{4.55}C$ sample using a Phi Electronics, Inc., Chanhassen, USA, Versaprobe II equipped with a monochromatic Al X-ray source with a 100 μm spot size at 25 W. The high-resolution scan of the boron B1s peak was taken with a 0.1 eV step size and a pass energy of 23.5 eV. Our XPS analysis shows that 64.7% of the boron is B-B bonded (binding energy of 187.32 eV), and the remaining 35.3% is B-C bonded (binding energy of 188.5 eV). This B-C binding energy is consistent with the value of 188.6 eV reported for chemical states of boron B1s reported for boron–carbide in the literature [13].

Figure 5 shows the Raman spectrum from the $B_{4.55}C$ sample in the 200–1300 wavenumbers range recorded with a laser wavelength of 532 nm. Several high-frequency modes in the range of 700 cm^{-1} and 900 cm^{-1} are attributed to intra-icosahedral or inter-icosahedral B-B bonds, while the strong Raman band at 1084 cm^{-1} is attributed to the breathing mode of the icosahedron [14]. It should be noted that, in comparison to the Raman spectrum of stoichiometric B_4C [14], Figure 5 shows additional Raman modes at 386 cm^{-1} and 1161 cm^{-1}. These additional Raman modes are characteristic of boron-rich boron–carbide materials, as 1161 cm^{-1} mode is thought to occur from a change in the breathing mode of icosahedron due to excess boron, and 386 cm^{-1} mode is attributed to boron substitution, leading to the formation of the C-B-B linear chain in a boron-rich boron–carbide material.

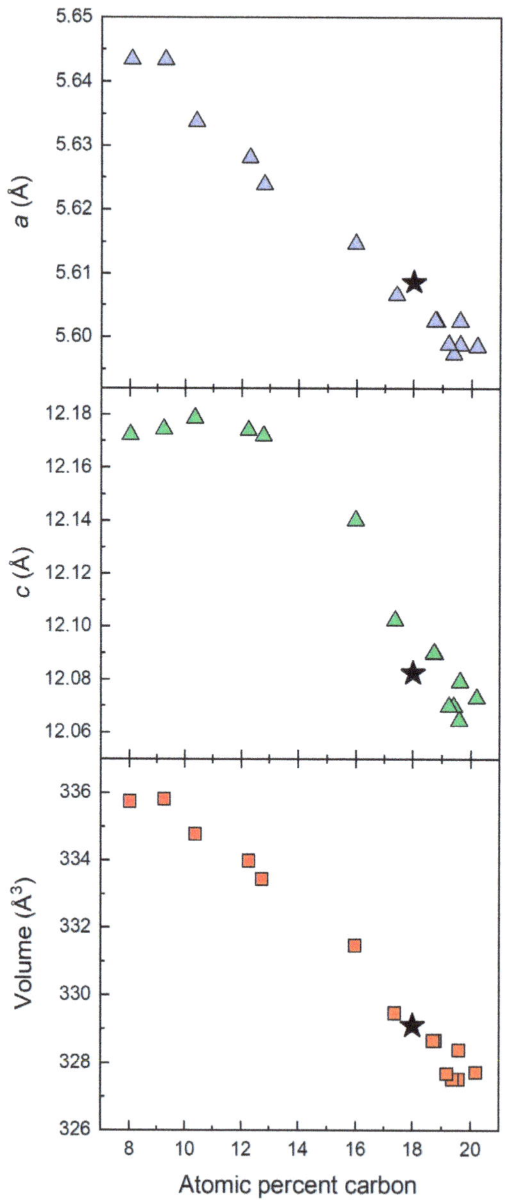

Figure 3. Our measured lattice parameters and unit cell volume for the rhombohedral phase of $B_{4+\delta}C$ in the hexagonal representation marked by asterisk * is superposed on data from Ref. [12] between atomic percent carbon range of 8.8% and 20%. This superposition results in 18 atomic percent carbon or material composition $B_{4.55}C$ in our case.

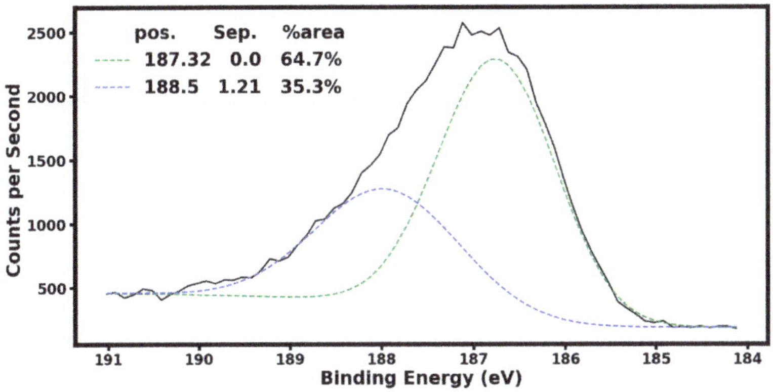

Figure 4. The measured X-ray photoelectron spectroscopy (XPS) high-resolution scan of the boron 1 s peak from the $B_{4.55}C$ sample showing a doublet structure with peaks at 188.5 eV (B-C bonding) and 187.32 eV (B-B) bonding.

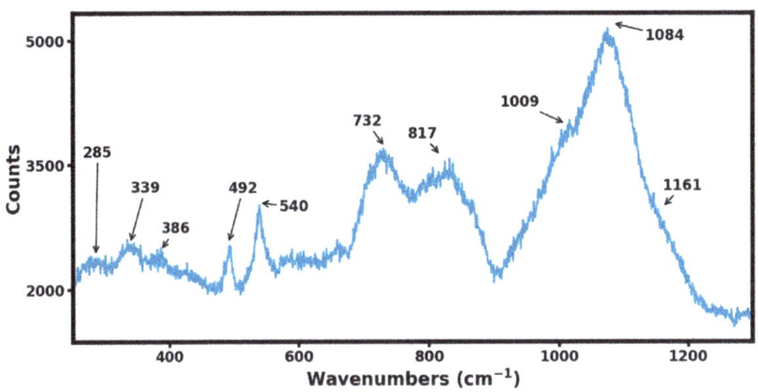

Figure 5. The Raman spectrum of the $B_{4.55}C$ sample recorded with a laser excitation of 532 nm. The Raman mode frequencies are indicated, and their assignments are discussed in the text.

The thermal oxidization resistance of the $B_{4.55}C$ sample was measured in compressed dry air at a temperature of 1300 °C. The previous high-temperature oxidation behavior of boron–carbide has been limited to a temperature of 1000 °C [15]. To have a precise comparison between various experimental runs on the $B_{4.55}C$ sample and a comparison with other high-temperature ceramics, all samples were sectioned into cubes so that their surface area could be measured, allowing the measured gain in mass to be normalized by surface area. Once the sample mass and surface area are recorded, the sample is loaded into the TGA furnace. Air is then evacuated from the furnace/balance chamber and then refilled with compressed dry air. Compressed dry air is slowly circulated through the furnace to replenish the oxygen supply. Figure 6 shows the results of two different experiments on $B_{4.55}C$ samples at 1300 °C labeled B-C Expt.1 (27.5 mg sample with a surface area of 33.27 mm^2) and B-C Expt.2 (18 mg sample with a surface area of 27.45 mm^2). The two experimental high-temperature runs are in reasonable agreement with each other, given the experimental uncertainties. Figure 6 also shows a similar thermal oxidation study at 1300 °C on a high-entropy boride (HEB) with a chemical composition of $(HfMoNbTaZr)B_{10}$. It is important to note that the thermal oxidation resistance of the $B_{4.55}C$ sample is better than that of high-temperature ceramics such as HEB, especially at temperatures approaching 1300 °C.

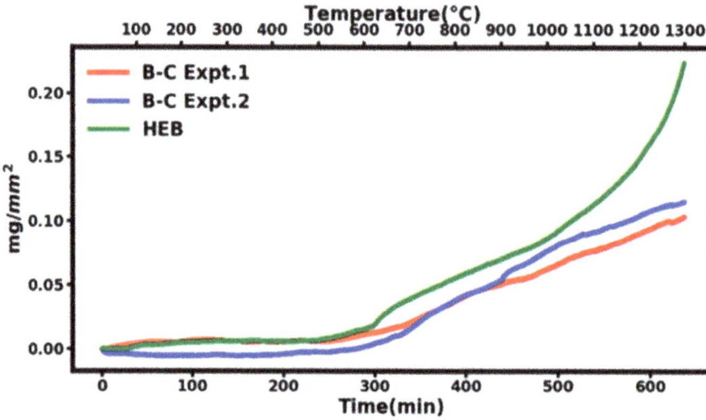

Figure 6. The measured increase in mass expressed as mg/mm^2 of the $B_{4.55}C$ sample and a high-entropy boride (HEB) as a function of temperature up to 1300 °C. Both samples were heated at a constant rate of 2° per min, and mass gain was normalized to surface area for a quantitative comparison. The zero curve with an empty crucible has been subtracted from all the data.

Vickers hardness measurements were performed using a Phase II+ 900-390A Micro Vickers Hardness tester (TEquipment, Long Branch, NJ, USA). The measured hardness data for the $B_{4.55}C$ sample at various loads with a dwell time of 15 s is shown in Table 1. Our measured value of hardness of 33.8 GPa at a maximum load of 200 gf is slightly higher than the 28–30 GPa values reported for various boron-rich boron–carbide materials [14] and could be related to differences in the microstructure of synthesized materials.

Table 1. Measured Vickers hardness for the $B_{4.55}C$ sample at various applied loads with dwell times of 15 s each.

Applied Load in g	Vickers Hardness (GPa)
50 gf	39.2 ± 5.1 GPa
100 gf	37.0 ± 0.8 GPa
200 gf	33.8 ± 2.3 GPa

4. Discussion

Our direct synthesis of boron-rich boron–carbide material using spark plasma synthesis at high pressure and high temperatures indicates the stability of a rhombohedral phase with a composition of $B_{4.55}C$ (18 atomic percent carbon). These results are consistent with the fact that a single rhombohedral phase is stable between the range of $B_{10.4}C$ (8.8 atomic percent carbon) and B_4C (20 atomic percent carbon). The X-ray photoelectron spectroscopy measurements reveal the characteristic binding energy of 188.5 eV for a boron–carbon bond established between the carbon and boron precursors during the high-pressure, high-temperature sintering process. The high hardness value of 33.8 GPa for $B_{4.55}C$ material is consistent with it being the third hardest material behind cubic boron nitride and diamond. Raman spectroscopy of $B_{4.55}C$ material documented the presence of an additional vibrational mode at 386 cm^{-1} that is attributed to excess boron that replaces carbon in the linear chain C-B-C to C-B-B. In addition, excess boron results in the distortion of the breathing mode of the boron icosahedron and results in a new Raman mode at 1161 cm^{-1}. The thermal oxidation studies in compressed dry air on $B_{4.55}C$ material show less mass gain per unit area at 1300 °C as compared to the high-entropy transition metal boride, establishing it as a prime candidate for applicability in extremes of pressure, temperatures, and oxidizing environments. Future studies should aim at extending the thermal oxidation resistance of boron-rich boron–carbide materials to higher temperatures of around 2000 °C.

5. Conclusions

We report on the direct synthesis of boron-rich boron–carbide $B_{4+\delta}C$ material using a ball-milled mixture of graphite and boron powders by spark plasma sintering at a pressure of 7 MPa and at a temperature of 1930 °C. The synthesized material is a rhombohedral phase with measured lattice parameters in hexagonal representation of $a = 5.609 \pm 0.007$ Å and $c = 12.082 \pm 0.02$ Å, and comparison with published trends shows the composition to be $B_{4.55}C$. X-ray photoelectron spectroscopy reveals a B-C bond with a binding energy of 188.5 eV, and Vickers hardness yields a high hardness value of 33.8 GPa. Raman spectroscopy has identified a new vibrational mode at 386 cm^{-1} representative of C-B-B linear chain formation due to excess boron in the lattice. There is an additional vibrational mode at 1161 cm^{-1} that can be attributed to the distortion of the breathing mode associated with the boron icosahedron. The quantitative thermal oxidation in compressed dry air indicates that $B_{4.55}C$ material has a better oxidation resistance behavior at 1300 °C compared to high-temperature ceramics such as the high-entropy transition metal borides. The high neutron absorption cross section, high melting point, high mechanical strength, and thermal oxidation resistance make the boron-rich boron–carbide material ideally suited for applications in extreme environments.

Author Contributions: Y.K.V. conceived the idea of direct synthesis of boron-rich boron–carbide material by spark plasma synthesis and supervised all the experimental studies reported in this paper. S.I. and W.S. contributed to the synthesis and characterization of the materials. R.S. contributed to X-ray structural refinements. P.A.B. contributed to XPS and XRD measurements. All authors contributed to writing and editing their specific portions of the manuscript. All authors have read and agreed to the published version of the manuscript.

Funding: This material is based upon work supported by the Department of Energy-National Nuclear Security Administration under Award Number DE-NA0004090. W.S. acknowledges support from the National Science Foundation (NSF)—Research Experiences for Undergraduates (REU)-site award to UAB under Grant # DMR-2148897. The spark plasma sintering facility at UAB was acquired under Grant # 2116564, awarded by the NSF Major Research Instrumentation program.

Data Availability Statement: All data generated or analyzed, and materials synthesized during this study are included in this published article. Additional raw data used and/or analyzed during the current study are available from the corresponding author on reasonable request.

Conflicts of Interest: The authors declare no conflict of interest.

References

1. Oganov, A.R.; Chen, J.; Gatti, C.; Ma, Y.; Ma, Y.; Glass, C.W.; Liu, Z.; Yu, T.; Kurakevych, O.O.; Solozhenko, V.L. Ionic high-pressure form of elemental boron. *Nature* **2009**, *457*, 863–867. [CrossRef] [PubMed]
2. Shirai, K. Phase diagram of boron crystals. *Jpn. J. Appl. Phys.* **2017**, *56*, 05FA06. [CrossRef]
3. Terekhova, A.M.; Levchenko, Y.V.; Matveeva, T.V.; Borisov, I.S.; Chelmakov, I.A. Influence of boron enrichment in control rods on neutron—Physical characteristics of the core of the reactor VVER. *IOP Conf. Ser. Mater. Sci. Eng.* **2019**, *487*, 012007. [CrossRef]
4. L'Annunziata, M.F. Chapter 1—radioactivity and our well-being. In *Radioactivity*, 2nd ed.; Elsevier: Boston, MA, USA, 2016; pp. 1–66.
5. Longuet-Higgins, H.C.; Roberts, M.D.V. The electronic structure of an icosahedron of boron atoms. *Proc. R. Soc. Lond. Ser. A Math. Phys. Sci.* **1955**, *230*, 110–119.
6. Sanz, D.N.; Loubeyre, P.; Mezouar, M. Equation of state and pressure induced amorphization of β-boron from X-ray measurements up to 100 Gpa. *Phys. Rev. Lett.* **2002**, *89*, 245501. [CrossRef] [PubMed]
7. Laubengayer, A.W.; Hurd, D.T.; Newkirk, A.E.; Hoard, J.L. Boron. i. preparation and properties of pure crystalline boron. *J. Am. Chem. Soc.* **1943**, *65*, 1924–1931. [CrossRef]
8. Will, G.; Kossobutzki, K. X-ray diffraction analysis of B50C2 and B50N2 crystallizing in the 'tetragonal' boron lattice. *J. Less-Common Met.* **1976**, *47*, 33–38. [CrossRef]
9. Baker, P.A.; Chen, W.; Chen, C.-C.; Catledge, S.A.; Vohra, Y.K. First-Principles Predictions and Synthesis of B50C2 by Chemical Vapor Deposition. *Sci. Rep.* **2020**, *10*, 4454. [PubMed]
10. Jay, A.; Duparc, O.H.; Sjakste, J.; Vast, N. Theoretical phase diagram of boron carbide from ambient to high pressure and temperature. *J. Appl. Phys.* **2019**, *125*, 185902. [CrossRef]

11. Rasim, K.; Ramlau, R.; Leithe-Jasper, A.; Mori, T.; Burkhardt, U.; Borrmann, H.; Schnelle, W.; Carbogno, C.; Scheffler, M.; Grin, Y. New facets of the known material—Local atomic arrangements and band structure of boron carbide. *Angew. Chem. Int. Ed.* **2018**, *57*, 6130–6135. [CrossRef] [PubMed]
12. Aselage, T.; Tissot, R.G. Lattice constants of Boron Carbides. *J. Am. Ceram. Soc.* **1992**, *75*, 2207–2212. [CrossRef]
13. Jacobsohn, L.G.; Schulze, R.K.; Maia da Costa, M.E.H.; Nastasi, M. X-ray photoelectron spectroscopy investigation of boron carbide films deposited by sputtering. *Surf. Sci.* **2004**, *572*, 418–424. [CrossRef]
14. Cheng, C.; Reddy, K.M.; Hirata, A.; Fujita, T.; Chen, M. Structure and mechanical properties of boron-rich boron carbides. *J. Eur. Ceram. Soc.* **2017**, *37*, 4514–4523. [CrossRef]
15. Dokumacı, E.; Ozkan, I.; Bülent Onay, A. High-temperature oxidation behavior of boron carbide. *J. Boron.* **2018**, *3*, 151–156. [CrossRef]

Disclaimer/Publisher's Note: The statements, opinions and data contained in all publications are solely those of the individual author(s) and contributor(s) and not of MDPI and/or the editor(s). MDPI and/or the editor(s) disclaim responsibility for any injury to people or property resulting from any ideas, methods, instructions or products referred to in the content.

materials

MDPI

Review

Phase Transitions in Boron Carbide

Helmut Werheit

Faculty of Physics, University Duisburg-Essen, D-47048 Duisburg, Germany; helmut.werheit@koeln.de

Abstract: The idealized rhombohedral unit cell of boron carbide is formed by a 12-atom icosahedron and a 3-atom linear chain. Phase transitions are second order and caused by the exchange of B and C sites or by vacancies in the structure. Nevertheless, the impact of such minimal structural changes on the properties can be significant. As the X-ray scattering cross sections of B and C isotopes are very similar, the capability of X-ray fine structure investigation is substantially restricted. Phonon spectroscopy helps close this gap as the frequency and strength of phonons sensitively depend on the bonding force and mass of the vibrating atoms concerned. Phase transitions known to date have been identified due to significant changes of properties: (1) The phase transition near the chemical composition B_8C by clear change of the electronic structure; (2) the endothermic temperature-dependent phase transition at 712 K according to the change of specific heat; (3) the high-pressure phase transition at 33.2 GPa by the drastic change of optical appearance from opacity to transparency. These phase transitions affect IR- and Raman-active phonons and other solid-state properties. The phase transitions at $B_{\sim 8}C$ and 712 K mean that a well-defined distorted structure is converted into another one. In the high-pressure phase transition, an apparently well-defined distorted structure changes into a highly ordered one. In all these cases, the distribution of polar C atoms in the icosahedra plays a crucial role.

Keywords: boron carbide; phase transition; structural disorder; electronic properties; phonons

1. Introduction

Boron carbide is one of the technically most important boron-rich solids. Outstanding properties such as low density (2.52 g/cm^3), high melting point (T_m~2700 K), and extreme hardness (~40,000 MPa) predestine boron carbide for application under conditions that are inaccessible for most other materials. Therefore, it is one of the best-investigated boron-rich solids where its structures are based on B_{12} or related icosahedra. The complexity of the structure and extraordinary electronic properties have attracted attention. Regarding technical application, boron carbide is the most important boron-rich material. Over the last century, its hardness and chemical stability have been utilized in powders for grinding and polishing, in nozzles for sandblasting, and in mortars (see A. Lipp [1]). Meanwhile, high-temperature ceramics and lightweight armoring of objects and persons have become important. High neutron absorption of the ^{10}B isotope allows application for neutron shielding and control rods in nuclear devices. Domnich et al. [2] reviewed such applications. In general, due to its extraordinary properties, boron carbide is highly promising for application under extreme conditions, such as high temperatures and mechanical or chemical burdens. Mostly in such applications, details of the complex crystalline structure of boron carbide, such as the distribution of C atoms, are of subordinate significance.

This does not apply in the case of the electronic properties. Optical appearance and electronic transport properties characterize boron carbide clearly as a p-type semiconductor (see [3,4]). However, untypical behavior indicates the strong impact of structural defects.

The high Seebeck coefficient (S~300 mVK^{-1} up to at least 2000 K) is a promising potential for direct thermoelectric energy conversion (see [5]).

Further uncommon effects, such as low-T specific heat anomaly [6] and the effects of fast neutron irradiation on the thermal conductivity [7], might also be associated with structural changes.

At least concerning details, the crystalline structure of boron carbide has not yet been finally determined. The formerly frequently and even currently occasionally assumed representative compound B_4C (structure formula ($B_{11}C$) CBC) does not exist in reality. $B_{4.3}C$ is the real compound achieved in commercially produced boron carbide [8,9] and in single crystals [10]. If B_4C is obtained, excess carbon is precipitated in the structure, typically in the form of graphitic layers [11,12]. Unlike $B_{4.3}C$ at the carbon-rich limit, the boron-rich limit of the homogeneity range is less well-defined. A few samples, whose C content was clearly below $B_{11}C$, were unambiguously characterized as boron carbide, albeit with a large error margin of lattice parameters (see [13] and references cited therein). Otherwise, β-rhombohedral boron with clearly higher C content has been obtained (see [14] and references cited therein). Therefore, the coexistence of both structures in a certain range cannot be excluded. Theoretical studies claim the structure model (B_{12}) CBC to be the energetically most favorable, while various experiments have proved this compound to be the most distorted one in the whole homogeneity range (see Reference [3]).

Phase transitions, structural changes related, and their effect on some other solid-state properties are the object of this paper.

2. Structure

Figure 1 [13] shows the rhombohedral unit cells of α-rhombohedral (α-rh. B; idealized structure formula B_{12}) boron and boron carbide (idealized structure formula (B_{12}) CBC and ($B_{11}C$) CBC), respectively. The actual structures differ more or less decisively—in α-rhombohedral boron, IR phonon spectra suggest the existence of additional single B atoms [15]. In boron carbide, where its homogeneity range extends from $B_{4.3}C$ at the carbon-rich to about $B_{11}C$ at the boron-rich limit, high-density structural defects have been identified and impact the electronic properties decisively [16,17].

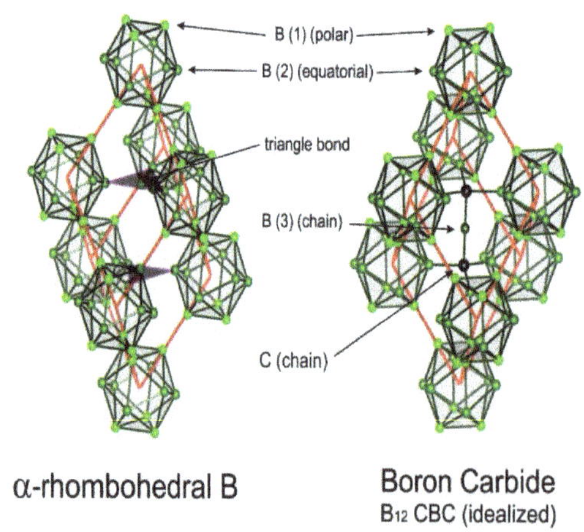

Figure 1. Idealized rhombohedral unit cells of α-rhombohedral (structure formula B_{12}) boron and boron carbide (structure formula B_{12} CBC) [12]. In ($B_{11}C$) CBC at the carbon-rich limit, the icosahedral C atom occupies one of the B(1) polar sites. Reprinted from Ref. [13] published by Elsevier Masson SAS. All rights reserved.

The defects in boron carbide are mainly formed by the exchange of B and C atoms on regular sites or of minor concentrations on unoccupied sites. X-ray diffraction averaging larger volumes yields the basic parameters of boron carbide reliably, but it is largely unsuitable to analyze individual compositions of the components. Due to the similar X-ray scattering cross sections of ^{10}B, ^{11}B, and ^{12}C isotopes, the capability of X-ray fine structure investigation is significantly restricted. Phonon spectroscopy helps close this gap.

Helpful in this respect is the comparison of the spectra of IR- and Raman-active phonons with those of α-rhombohedral boron having the same icosahedral structure; however, without C atoms on polar sites of the icosahedron and without CBC or these replacing other structure elements. Figure 2 compares the IR-phonon spectra of single crystal $^{nat}B_{4.3}C$ boron carbide and α-rhombohedral boron (for comparison of the Raman spectra, see References [18–20]). The spectrum of α-rhombohedral B was discussed in Reference [15]. A detailed theoretical description was performed by Beckel et al. [21]. The spectra of boron carbide were analyzed by Werheit et al. (see Refs. [13,22] and the references cited therein). The concentration of structure elements depending on the C content is shown in Figure 3. Changes in some icosahedral vibrations in boron carbide in connection with phase transitions are discussed below.

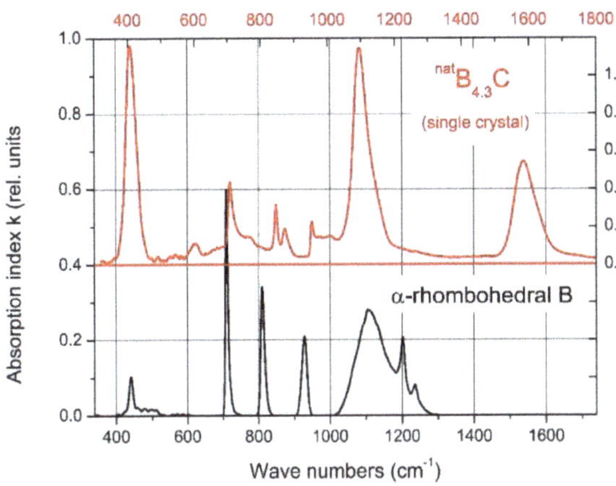

Figure 2. IR-phonon spectra of $B_{4.3}C$ boron carbide and α-rhombohedral boron. The abscissas are shifted relative to one another to facilitate the correlation of icosahedral vibrations.

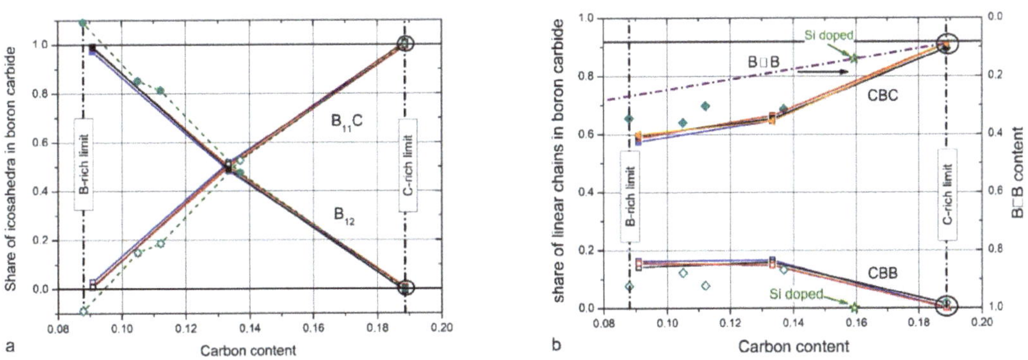

Figure 3. Shares of structure elements in the unit cell of boron carbide [16,22]; (**a**), B_{12} and $B_{11}C$ icosahedra, (**b**), CBC, CBB, and B□B chains (□, vacancy). Symbols connected with solid lines, isotope-enriched boron carbide; other symbols, polycrystalline boron carbide. Reprinted from Ref. [22] "© IOP Publishing. Reproduced with permission. All rights reserved".

The existence of the following defects in the structure in the form of deviations from the idealized structure formulas is generally confirmed:

- C atom replacing B(1) in ($B_{11}C$) CBC; that means one of the six B atoms on polar sites of the icosahedra is replaced by a C atom. The distribution of these C atoms is an object of current investigation.
- Carbon-rich limit of the homogeneity range: $B_{4.3}C$ (instead of B_4C, formerly assumed to be the actual compound of boron carbide) [6,7,14]. Hence, the real structure contains 5.7 at. % structural defects compared with the idealized ($B_{11}C$) CBC structure. This also applies to high-quality single crystals, whose preparation is only possible in the chemical composition $B_{4.3}C$ [8], and to single crystals growing accidentally in industrial production (concluded from the electrical conductivity, see below Ref. [23]).
- Vacancies of the B(3) site (9–28%, depending on composition) (see [14] and references cited therein).
- CBB chains replace CBC with boron-rich boron carbides (Figure 3b, see [14] and references cited therein).

$B_{4.3}C$ is the least distorted compound [14,15]. It does not contain any free carbon, for example, systematically checked by Schwetz et al. [8,24] using a wet-chemical analysis for determining free and bound carbon in boron carbide separately. $B_{4.3}C$ is the only compound allowing the preparation of high-quality single crystals [8].

Theoretical model calculations determined the idealized structure formula (B_{12}) CBC of the compound $B_{13}C_2$ as the energetically most favorable [25,26]. In fact, $B_{13}C_2$ is the most distorted boron carbide. Its highly complex structure $(B_{12})_{0.5}(B_{11}C)_{0.5}\bullet(CBC)_{0.65}(CBB)_{0.16}(BB)_{0.19}$ shows that about 80% of the unit cells contain any structural defect [4,27,28]. High concentrations of structural defects (with regard to idealized structures) are a generic property of boron carbide structure.

The shares of structure elements in the unit cell depend on carbon content (see Figure 3 [22]), preferably from the investigation of IR- and Raman-active phonons. The comparably low share of B□B chains (□, vacancy) has not been measured but is used to close the gap between the sum of atoms in the other structure elements determined and the chemical composition. TEM studies by Rasim et al. [29] suggest that a variety of further kinds of structural elements can replace the CBC chain. However, apart from this TEM study, there is no further experimental evidence. Therefore, it cannot be excluded that such structural elements are provoked by the preparation process of the sample.

Many theoretical studies on structural details in boron carbide are based on Raman studies, which are affected by a severe systematic experimental error [18]. In cases of high Raman excitation energy, the spectra do not reflect bulk properties. Hence, the results of accordingly performed calculations merely end in themselves and do not yield scientific progress (see, for example, the controversial discussions in [19,30–34]).

3. Electronic Structure of $B_{4.3}C$ at Ambient Conditions

Due to the availability of material, previous experimental studies of electronic properties are mainly restricted to polycrystalline boron carbide and especially to the compound $B_{4.3}C$. This also applies when the compound is nominally designed as B_4C, as this is, in reality, $B_{4.3}C$ with excess carbon typically distributed in the form of graphitic layers [11,24]. High-quality $B_{4.3}C$ single crystals were used in References [10,35,36]. Figure 4 shows a collection of transition energies, consistently obtained from optical absorption such as those shown in Figure 5, indicating interband transitions at 2.09 and 2.41 eV and from electrical transport measurements (see [37–41] and references cited therein).

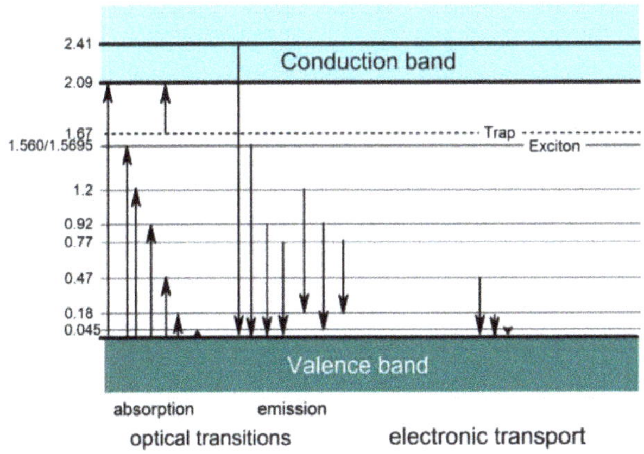

Figure 4. Empirical band scheme of boron carbide based on experimental results (see [3,4] and references cited therein).

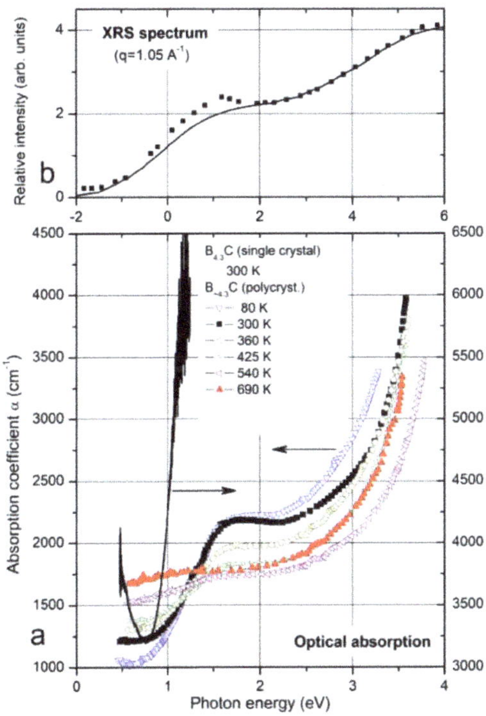

Figure 5. $B_{4.3}C$. (**a**) Optical absorption edge [3,4,42]. Symbols, polycrystalline material; line, single crystal. (**b**) X-ray Raman scattering (XRS) spectrum [43]. Symbols selected experimental data for q = 1.05 °A^{-1}; line, site-specific ab initio calculation for the background of icosahedral B atoms. We assume that the composition of this boron carbide corresponds to $B_{4.3}C$, the carbon-rich limit of the homogeneity range; the composition B_4C claimed by the authors does not exist (see above).

Off the absorption edge in Figure 5, there is a peak absorption in polycrystalline boron carbide at 1.56 eV. On the first view, surprisingly, the absorption of high-quality single-crystal boron carbide in this range is considerably higher. XRS emission (Figure 5,

Feng et al. [43]) and luminescence [10,42] (Figure 6) suggest a p-type exciton associated with the central B atom of the CBC chain. The luminescence peak consists of two components, where the intensity relation is 4:1. This corresponds quite well with the relation in $^{nat}B_{4.3}C$ $^{11}B/^{10}B = 4.03$ if random isotope distribution is assumed. The formation of excitons requires largely undistorted structures. In $B_{4.3}C$ single crystals, this condition is better fulfilled than in polycrystals, and at low T, it is better than at high T. This explains the peculiarities in Figure 5. Hence, the gap state "Exciton" in Figure 4 is not evoked by structural defects. However, the effects of structural defects on the optical absorption cannot be excluded [44]. For example, at high pressure (see below), structural distortion occurs by buckling out of the central B atom in the CBC chain [15,23,35,36].

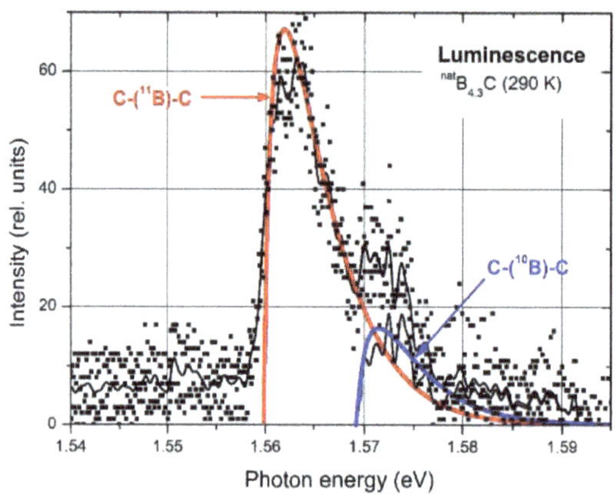

Figure 6. $B_{4.3}C$, photoluminescence spectrum at 290 K [42]. Excitation with the 514.5 nm (2.4 eV) line of an Ar laser; intensity 280 mW mm^{-2}. Squares, experimental results; thin black lines, averaged experimental results, before and after subtracting the 1.56 eV model fit, respectively; thick colored lines, recombination models of free excitons (1.560 and 1.5695 eV, respectively).

Former theoretical studies on boron carbide were not very helpful in understanding the complex interactions between structural details and electronic properties [45–51], as they were based on idealized structure models deviating more or less considerably from the real structures. Thus, related conclusions were often in significant contrast to experimental results. For example, the theoretically claimed metallic character or even superconductivity [48] was opposite to the semiconducting behavior experimentally proved from the very beginning. Surprisingly, even in recent studies, the real structure of boron carbide is frequently ignored.

4. Phase Transition Near the Compound B_8C

Reliable information on the electronic structure of more boron-rich boron carbide is obtained from electrical conductivity, measured between 5 and 2100 K by Werheit et al. [37], by Wood [38,39] and by Aselage et al. [40]. Results obtained by Amulele et al. [41] up to 9 GPa pressure are compatible.

The activation energies obtained from the electrical conductivity are displayed in Figure 7, showing that at ~B_8C, the homogeneity range is divided into two parts with significantly different electronic characteristics. The activation energy at about 80 meV found in both regions is uncertain as obtained in that range of temperature, where a continuous transition from thermal excitation to variable-range hopping takes place (see [37]).

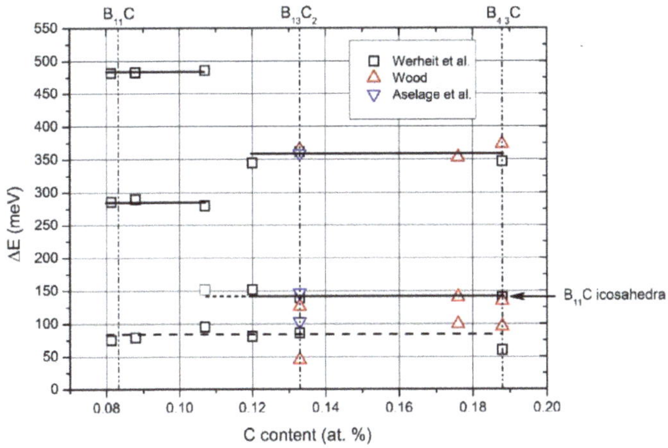

Figure 7. Phase transition near the compound B_8C.

It is expected that this phase transition is correlated with a structural change. This should be reflected in the phonon spectra. Figure 8 shows the IR- and Raman-active phonons of $^{11}B_xC$ boron carbide [52,53]. We analyze some phonons showing changes depending on the chemical composition.

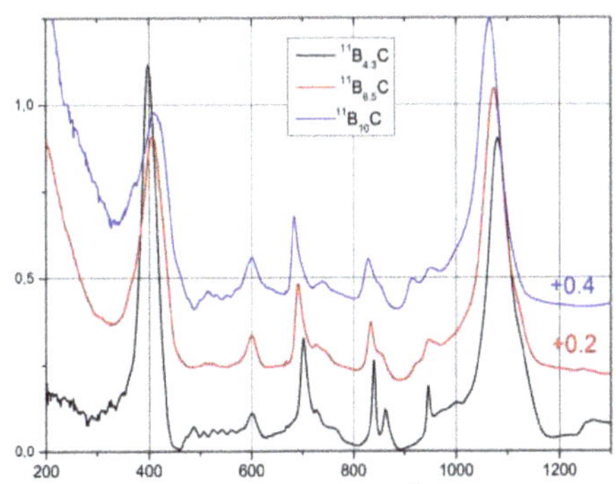

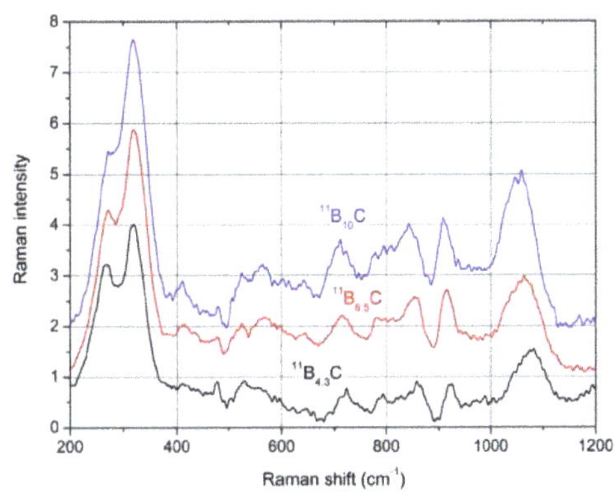

Figure 8. Phonon spectra of $^{11}B_xC$ at 30K [33,40]. (**a**), IR-active phonons; (**b**), Raman-active phonons.

On the basis of specific phonons, we check below how far structural changes are involved in the phase transitions. In most cases, the phonons concerned are no single modes but overlap in the spectrum. For separating the components, we used the "OriginPro software for graphic and analysis," allowing us to fit overlapping Gauss-like or Lorentz-like resonances. The phonons, which are affected by passing the phase transition, cannot be precisely attributed to concrete atoms or atomic arrangement. Accordingly, model calculations, which could help clear this deficiency, are missing so far.

4.1. IR-Active Phonon Near 400 cm^{-1}

The phonon near 400 cm^{-1} is composed of two components (see Figure 9a), where the frequencies (Figure 9b) and phonon strengths vary depending on the C content. Shirai et al. [54,55] and Vast et al. [50,56] attribute the IR phonon mode near 400 cm^{-1} to the central B atom in the chain vibrating perpendicularly to the chain axis [27]. Experimentally, this is confirmed for the component Ph 1, whose strength, depending on the C content, precisely matches the share of CBC chains (Figure 9b). The considerably weaker component Ph 2 of this phonon belongs to the icosahedral structure, accordingly with its counterpart in the spectrum of α-rhombohedral boron (see Figure 2). This phonon was already mentioned by Beckel et al. [21] but not described in detail (see [16]). Below, we perform an analysis of the temperature dependence of this phonon.

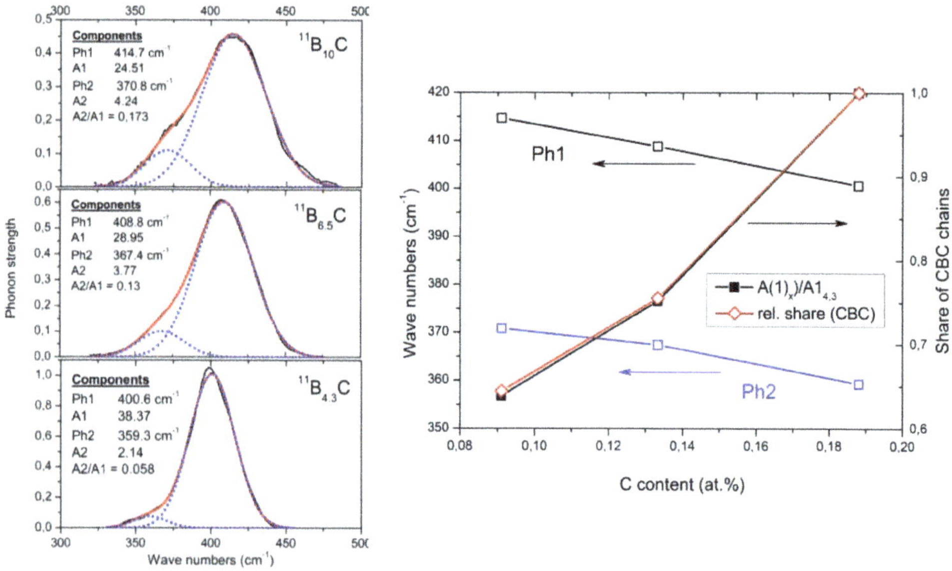

(**a**) (**b**)

Figure 9. (**a**). IR-active phonon of $^{11}B_xC$ near 400 cm^{-1} (All fits were obtained using the Origin Pro software). (**b**). Phonon shift depending on C content and comparison between the reduction of phonon strength (components 1 and 2) and relation of the phonon strength of component #1 depending on C content compared with the relative share of CBC chains in the structure (obtained from the IR-active stretching vibration near 1600 cm^{-1} in Figure 2b).

In Figure 9, no apparent change suggests a relation of this phonon to the phase transition at $B_{8.1}C$. However, the mode Grüneisen parameter γ of the IR-phonon (−3.21 at low and +6.5 at high temperatures) deviating strikingly from all other phonons [36] suggests a peculiarity. The previously assumed involvement of this phonon in the T-dependent transition at 712 K [57] is checked below.

4.2. *IR-Active Phonons Near 700 cm^{-1}*

The spectra of isotope-enriched samples (Figure 10a) exhibit clear changes depending on C content. For interpretation, a denser sequence of spectra is required. Data on $^{nat}B_xC$ are available ([57] Figure 10b). The results of the analysis (examples in (Figure 11a)) are displayed in Figure 11b. Splitting of the component of this phonon near 760 cm^{-1} into three branches indicates a structural change at the phase transition. This phonon is correlated with the strong 705 cm^{-1} intra-icosahedral single-line A_{2u} mode in α-rhombohedral boron, described by Beckel et al. [21], split in the boron carbide structure (see [16,31]).

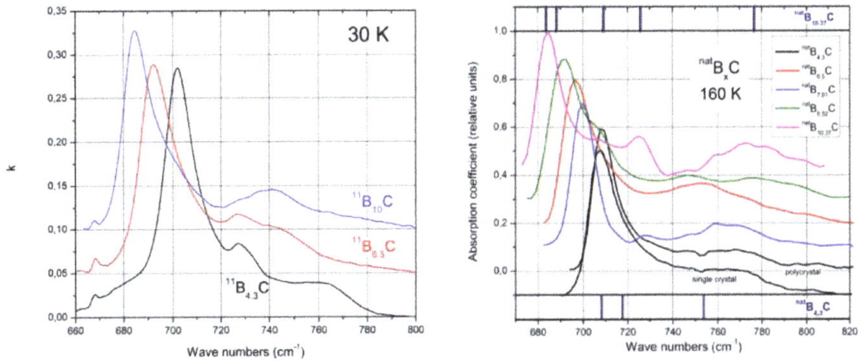

(a) (b)

Figure 10. IR-active phonons near 700 cm^{-1}. (**a**), $^{11}B_xC$ at 30 K; (**b**), $^{nat}B_xC$ at 160 K, phonon-wavenumbers of $^{nat}B_{4.3}C$ and $^{nat}B_{10.37}C$.

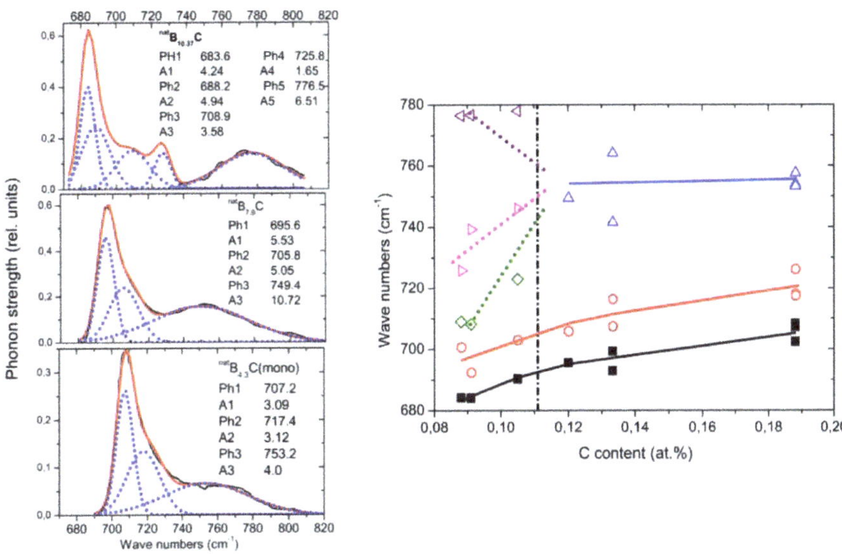

(a) (b)

Figure 11. IR-active phonons near 700 cm^{-1}. (**a**) Fits obtained with OriginPro software. (**b**) Wave numbers of phonon components vs. C content. The results of the high-frequency components of $B_{8.52}C$ (C content 0.105) are rather uncertain.

4.3. IR-Active Phonons Near 940 cm^{-1}

The phonon near 940 cm^{-1} (Figure 12) represents an icosahedral vibration proved by the counterpart in the spectrum of α-rhombohedral boron (see Figure 2). According to Beckel et al. [21], this phonon consists of superimposing E_u and A_{2u} modes. In boron carbide, the additional effect of polar C atoms in the icosahedra must be considered. This explains, in principle, the complexity of the phonon and of its change depending on C content.

However, there is a significant change between $B_{4.3}C$ and $B_{6.5}C$: two components of the phonon vanish completely. Unfortunately, there are no spectra in the intermediate range available, which could help explain this behavior.

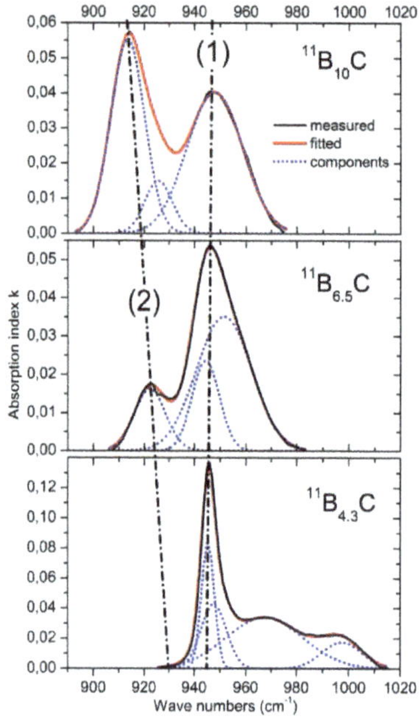

Figure 12. IR-active phonons near 950 cm^{-1}. (1) and (2) mark the apparent shift and strength of specific components.

4.4. FIR-Absorption

As indicated in the low-frequency region of Figure 8a, the FIR optical absorption increases towards very low wave numbers. Figure 13 shows this range extended down to 10 cm^{-1}. This absorption is caused by the plasma-like absorption of free charge carriers, which is heavily damped in boron carbide (for details, see [28]). With respect to the electric structure, Figure 13 shows that the concentration of free carriers increases with the C content decreasing. However, there is no effect of the phase transition at $B_{\sim 8}C$ on the concentration of free holes discernable.

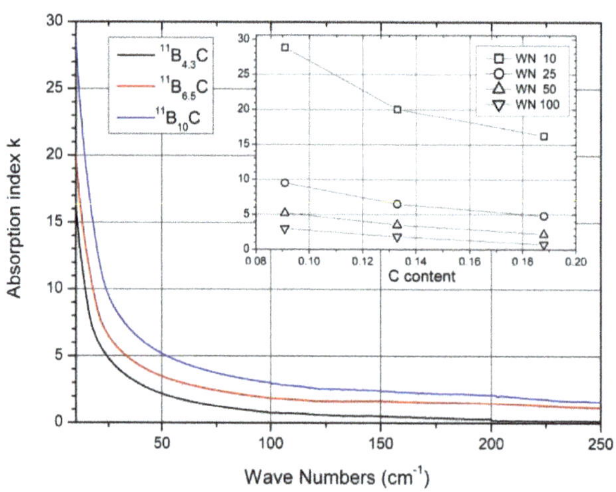

Figure 13. Heavily damped plasma absorption in $^{11}B_xC$ at 30 K [28].

The share of $B_{11}C$ icosahedra decreases from ~1 in $B_{4.3}C$ to ~0.1 in $B_{10}C$ (see Figure 3). The main components (1) and (2) seem to vary continuously depending on the C content. There is no indication of a sudden change near B_8C. Therefore, no involvement of this phonon in the phase transition at $B_{\sim 8}C$ is discernible.

4.5. Raman-Active Phonons

The resolution of the Raman spectra in Figure 8b is restricted. Therefore, variations dependence on the C content cannot be doubtlessly identified, apart from the strong phonon doublet at 270/320 cm^{-1} (Figure 14). This has no counterpart in the spectrum of α-rhombohedral boron. Therefore, it is to be attributed to the structural changes in boron carbide by C atoms: $B_{11}C$ icosahedra, CBC, or CBB chains.

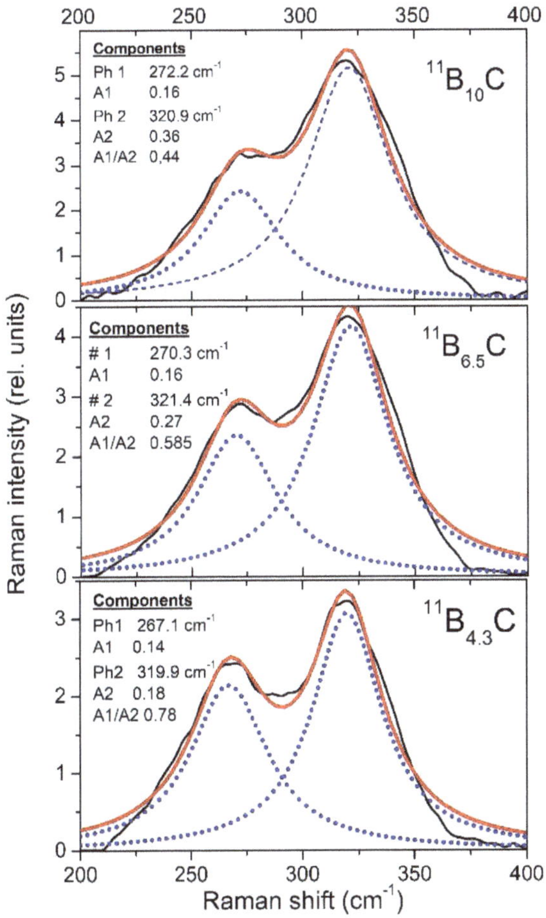

Figure 14. Raman-active phonons of $^{11}B_xC$ at 270/320 cm^{-1}.

We attribute this doublet to the E_g phonon at 335 cm^{-1} described in the theoretical study by Shirai and Emura [55] as a rotation of the CBC chain associated with a wagging icosahedron and assume that the splitting could result from the occupation of the atomic sites involved partially occupied by C or B atoms (see [12]).

The strengths of both components change monotonically depending on the carbon content. There is no indication that this phonon is involved in the phase transition at $B_{\sim 8}C$.

5. Phase Transition at 712 K

The T-dependent specific heat (Figure 15) shows an exothermic phase transition at 712 K, half-width 82 K [58]. The electrical conductivity proves the correlation of this phase transition with a structural change, where the activation energy of 140 meV is correlated with the content of the $B_{11}C$ icosahedra and polar C atoms, respectively (see above). At carbon contents below the phase transition $B_{\sim 8}C$, it is no longer detectable [50].

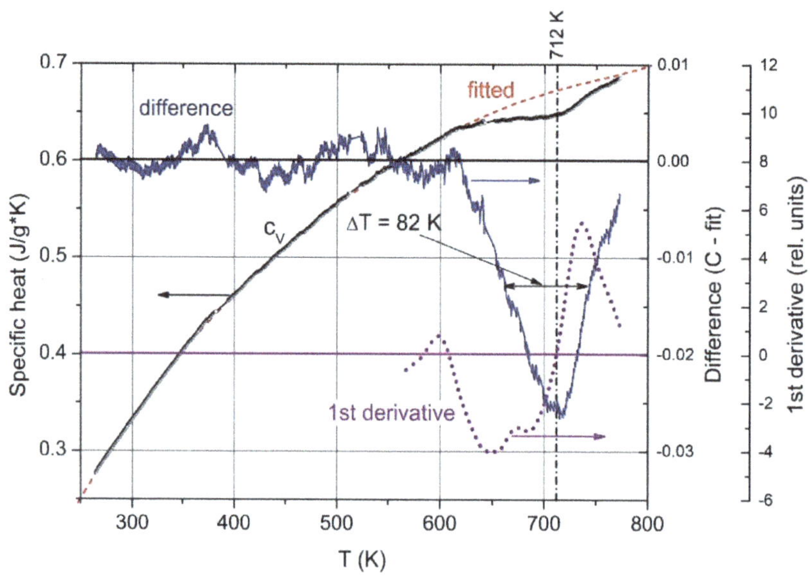

Figure 15. Specific heat of $B_{4.3}C$ boron carbide vs. T. Black circles, measured; red dashed line, polynomial fit; blue, difference between measured data and fitted line; purple dotted line, 1st derivative.

Yao et al. [59] described theoretically in this range an endothermic phase transition that is related to the site exchange of carbon atoms within the icosahedra. Obviously, this process is not compatible.

Complete temperature-dependent phonon spectra of boron carbide are available below 450 K only. The exception is the IR-active phonon mode near 400 cm^{-1}, available between 100 and 800 K (Figure 16a) [28]. The spectra suggested a structural change by mode splitting near the phase transition. However, this splitting is feigned.

As shown above, the IR-active phonon mode near 400 cm^{-1} consists of two components. One of them is an icosahedral vibration identified by the equivalent vibration in α-rhombohedral boron (see Figure 2). As such, it is mentioned by Beckel et al. [21] but not described in detail. Shirai et al. [54,55] and Vast et al. [51,56] attribute the second component to the central B atom in the CBC chain vibrating perpendicularly to the chain axis. We analyzed the IR absorption of this phonon accordingly. The parameters of both components are displayed in Figure 16b, showing that the before assumed split was feigned by the considerably different temperature dependence of the components.

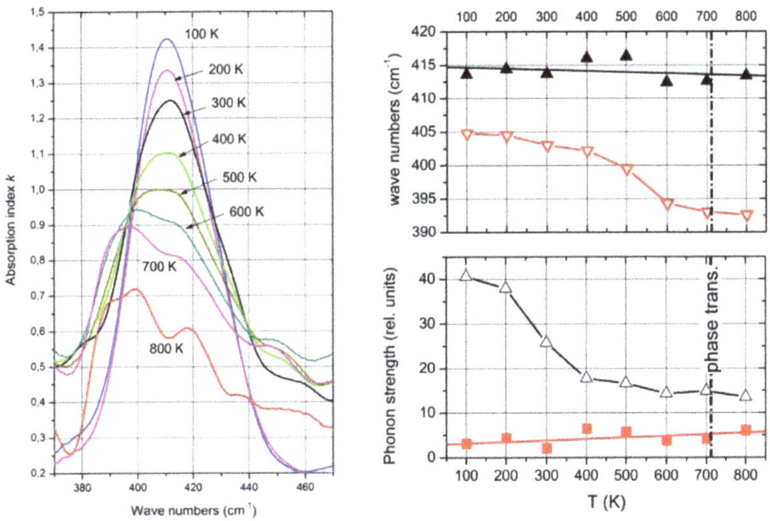

(a) (b)

Figure 16. IR-active phonon of ${}^{nat}B_{4.3}C$ boron carbide near 400 cm^{-1}. (**a**) Absorption index k vs. wave number between 100 and 800 K; (**b**) wave number and phonon strength of the components vs. T. Black, bending mode of the CBC chain; red, icosahedral vibration.

Passing the phase transition, the frequency of the component representing the icosahedral vibration decreases clearly from 405 cm^{-1} at low T to 395 cm^{-1}. Simultaneously, the phonon strength concerned increases marginally, showing that the number of vibrating elements remains nearly unchanged. The atomic bonding force concerned falls slightly in the range of phase transition but is nevertheless clearly measurable. It is surprising that the range of decrease is wider than the half-width of the phase transition (Figure 14). A certain effect of the phase transition on the electronic properties is clearly established.

The vibration of the central B atom in the chain behaves differently. The frequency remains nearly constant, showing that the bonds within the CBC chain are hardly affected by T. In contrast, the phonon strength representing the number of vibrating elements decreases considerably.

A certain difference between the structures below and above the phase transition is indicated by the frequency of the icosahedral phonon near 400 cm^{-1} (Figure 16b), which decreases from 405 cm^{-1} at low T to 392 cm^{-1} at high T. In contrast, the frequency of the bending mode of the central B atom in the CBC chain decreases at most marginally from ~414 to ~407 cm^{-1}. The redistribution of polar C atoms in the phase transition affects the bonding force within the icosahedron and, consequently, the electronic structure as well.

An important further characteristic of this phase transition can be obtained from the electrical conductivity at pressures up to 9 GPa [41]. The phase transition is reversible, however, with considerable relaxation time.

6. Phase Transition at 32 GPa

The drastic change from opacity to glasslike transparency is the visual appearance of the phase transition in boron carbide at 32 GPa. Hushur et al. [35] used the photos shown in Figure 17 to estimate the band gap increasing at increasing pressure roughly. This implies that the electronic structure of boron carbide known from ambient conditions [15] (see Figure 3) changes fundamentally at high pressure.

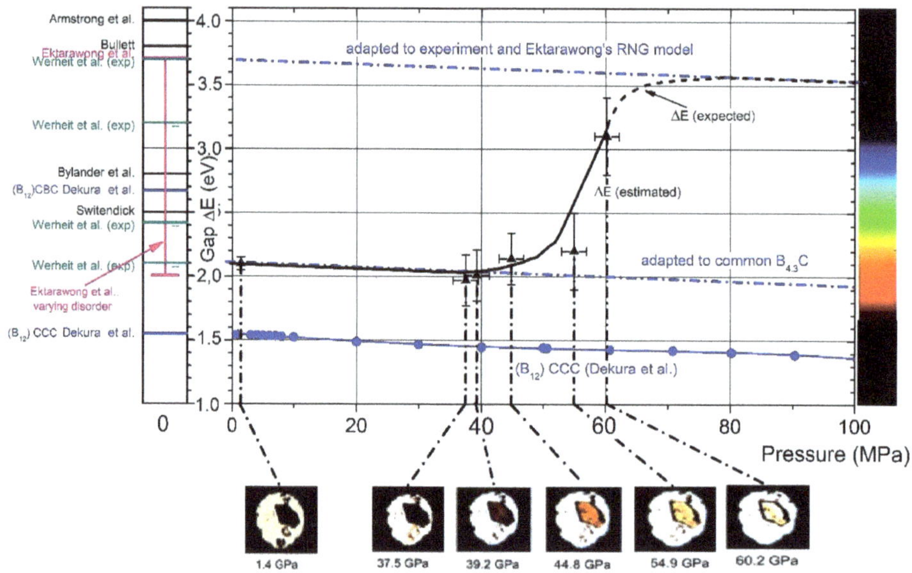

Figure 17. The gap width of boron carbide depending on pressure. Black triangles, gap width, roughly estimated from transient-light photos [35]. Compared with experimental and theoretical results [45,46,60,61]; pink, ordered structure [60]; pink vertical bar, varying configurational disorder [60]; blue filled circles, pressure-dependence [35]. Experimental results: cyan, ambient conditions [62,63]; dash-dotted lines, adapted to relevant gap widths at 0 GPa for distorted and undistorted $B_{4.3}C$ (Ektarawong's RNG model), respectively; pressure-dependence according to the calculation for $(B_{12})CCC$ [61].

In Figure 17 [35], the estimated shift of the band gap is compared with theoretical model calculations on undistorted and distorted boron carbides [45,46,60,61] and experimental results [44,62,63]. Obviously, this phase transition converts boron carbide, as it is commercially available or prepared at a laboratory scale and is highly distorted, into a largely undistorted state, which is at least close to the idealized models used in theoretical calculations.

At rising pressure, the band gap increases from values near 2 eV, experimentally determined [60] and theoretically attributed to a strongly distorted boron carbide [45] to gap widths between 3 and 4 eV, theoretically predicted for idealized meaning defect-free boron carbide.

Ektarawong et al. [45] studied how the distribution of icosahedral C atoms affects the electronic DOS. They used the idealized $(B_{11}C)$ CBC) structure and ignored changes in the CBC chain. In the icosahedra, six polar B(1) and six equatorial B(2) sites are occupied by B atoms, apart from one upper polar site p1 substituted by C. This C atom is slightly shifted towards the center of the icosahedron (see Reference [25] and references cited therein). The energetically most favorable defect structures are generated by the transfer of the C atom to both neighboring p2 and p3 sites, respectively, followed by transfers from p1 to the lower polar sites p4, p5, and p6, respectively. Ektarawong's RNG model is based on the assumption that the polar C atoms are randomly distributed over all six polar sites.

Rasim et al. [29] take the idealized (B_{12}) CBC structure model for reference. As mentioned above, this model is not compatible with real boron carbide. Nevertheless, this study excludes interfering effects in icosahedra when calculating the effects on the DOS by exchanging the CBC chain with other structure elements of various chain structures.

Both approaches mark significant achievements compared with the previous theoretical studies based on idealized and, therefore, unrealistic structure models. In Figure 18 [17], we check to what extent the DOS of both model calculations are compatible with the electronic transitions in the empirical band scheme (Figure 4) based on experimental results. These fit excellently with Ektarawong's RNG model, assuming a random distribution of icosahedral C atoms over all the polar sites [45].

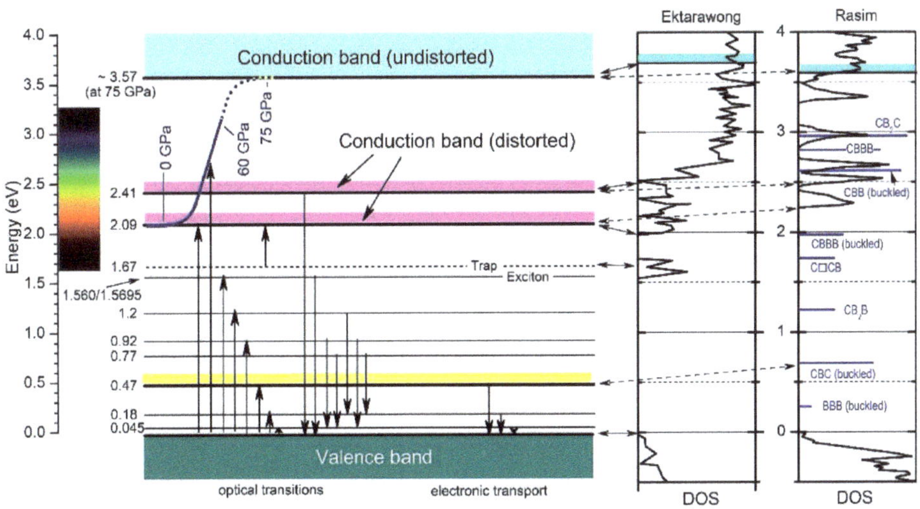

Figure 18. Electronic band scheme of $B_{4.3}C$ boron carbide [17] derived from optical and electrical measurements; DOS calculated by Ektarawong et al. (RNG model) [45] and Rasim et al. [29] for reference (data taken from diagram each). Vertical arrows show absorption (upward) and emission (downward) processes; the visible range is marked. Reprinted from Ref. [17] published by Elsevier Masson SAS. All rights reserved.

Considering these results, the strong optically excited transitions between ~2 and ~3.5 eV, which are responsible for the opacity of boron carbide at ambient pressure, are evoked in any way by the polar C atoms in boron carbide. The strong optical absorption concerned fulfills the requirements of electronic interband transitions and has been interpreted accordingly. Compared with classical semiconductors, the polar C atoms in boron carbide differ decisively from dopants having one excess electron compared with the basic lattice atoms and form low-density states in the gap. In boron carbide, the distribution of polar C atoms at room temperature induces high-density electronic states. A "simple" redistribution of the polar C atoms at high pressure, apparently to a higher degree of order, makes these high-density states completely disappear. This redistribution is reversible. That means the largely defect-free structure at high pressure [43] is metastable only, while the distorted structure is the thermodynamically stable one, at least under ambient conditions.

The reversibility of the structural distortion induced by the C atoms randomly distributed over all six polar sites stands in fundamental contrast to a sometimes-used model based on the assumption that the actual distorted structure of boron carbide is a non-equilibrium state only [28].

The DOS of boron carbide affected by structural elements replacing the CBC chain, calculated by Rasim et al. [29], is obviously not compatible with the high-pressure phase transition. Indeed, some experimentally determined narrow levels in the band scheme may correlate with specific structural elements assumed to replace the CBC chain. However, the related optical absorption is far too small to explain the strong absorption in this range at ambient conditions.

Therefore, the rearrangement of polar C atoms means a structural change and should be verifiable in the phonon spectra. The whole structure is affected by the phase transition at 33 GPa. This is proved by the phonon mode Grüneisen parameters γ. In the case of all IR- and Raman-active phonons, γ differs more or less considerably at pressure below and above the phase transition [13,30,36]. Unfortunately, in the pressure-dependent IR-phonon spectra of Chuvashova et al. [30], accordingly commented in [31] (some results are shown in Figure 19), the IR-active phonon near 400 cm^{-1} has been neglected. As shown above, a component of this phonon is sensitive regarding polar C atoms in $B_{11}C$ icosahedra.

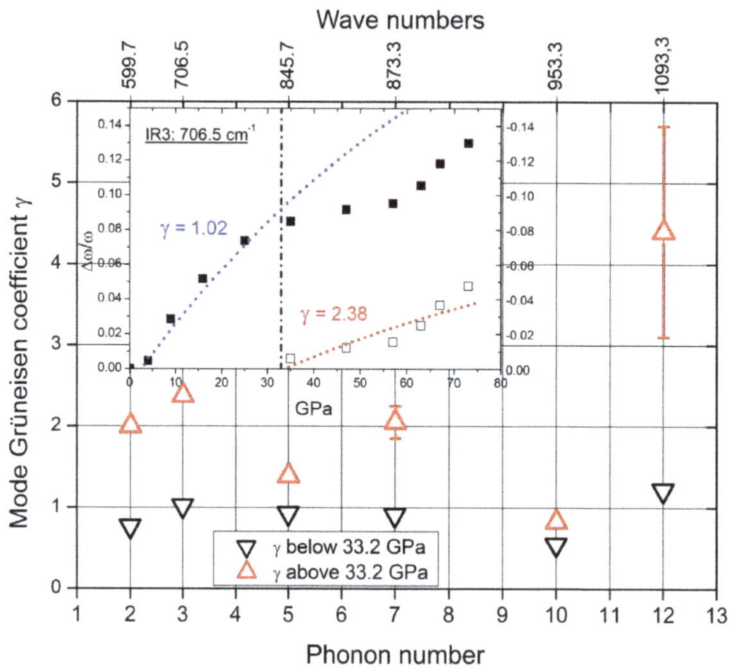

Figure 19. IR phonon mode Grüneisen parameters below (▽) and above (△) the phase transition at 33.2 GPa. Insert: an example of determining γ.

The spectra of IR- and Raman-active phonons (Figure 8) exhibit strong, probably partially split peaks near 1100 cm^{-1}. These are at least partially attributed to intra- and inter-icosahedral vibrations of polar atoms in the icosahedra. Hence, they should be affected by the rearrangement of polar C atoms.

In Figure 19, the Grüneisen coefficient of the 1093.3 cm^{-1} IR phonon above the phase transition considerably exceeds that of the other phonons. Moreover, the difference in the value below the phase transition is especially large. This suggests that the redistribution of polar C atoms increases the bonding force considerably.

The Grüneisen coefficients concerning Raman-active phonons, determined in Figure 20, change significantly at the phase transition due to rearranging polar C atoms.

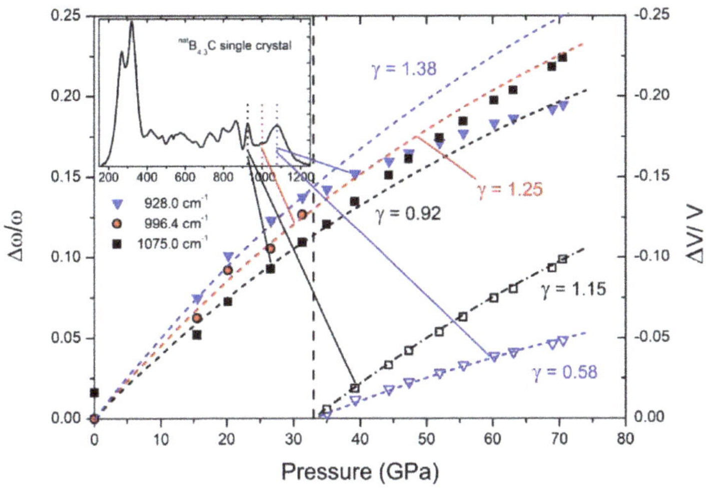

Figure 20. Relative shift of phonon frequencies (left ordinate) and unit cell volume (right ordinate) vs. pressure, determining mode γ of typical IR-active phonons of $B_{4.3}C$ boron carbide [13]. Insert: FT-Raman spectrum of single-crystal $B_{4.3}C$ boron carbide at ambient conditions. Reprinted from Ref. [13] published by Elsevier Masson SAS. All rights reserved.

In the Raman spectra of boron carbide, the phonon doublet at 270/320 cm^{-1} is the most striking feature. Werheit et al. [12] assigned this doublet to the 335 cm^{-1} E_g phonon that Shirai and Emura [55] attributed to a rotation of the 3-atom chain combined with a wagging icosahedron. Firstly, it is surprising that this phonon splits, though there are only $B_{11}C$ icosahedra in $B_{4.3}C$. This could be explained as follows. The translation vectors of atoms involved in this phonon (Figure 1 in Reference [53]) show that 8 of 12 atoms in the icosahedra and 4 of 6 polar sites, respectively, are involved. Assuming that the C atom is randomly distributed over all polar sites of the icosahedra, there are $B_{11}C$ icosahedra with C atoms on sites that are involved in this vibration, while others contain exclusively B atoms on such sites. Due to different m_{red}, the phonon mode splits and the components Ra 1 (~270 cm^{-1}; C on one of the relevant sites) and Ra 2 (~320 cm^{-1}; B atoms only) could be attributed accordingly.

The negative Grüneisen coefficients of both components differ fundamentally from all other phonons. Normally, compression reduces the lattice parameters, leading to a positive Grüneisen parameter, at least when the force constant is unaffected. However, that seems to be the case in boron carbide in connection with the rearrangement of polar C atoms at the high-pressure phase transition.

In Figure 21, some parameters of both components of this doublet are displayed. For the evaluation, it had to be considered that the spectra were obtained using a laser with problematic excitation energy (for details, see [17]). While the phonon frequencies are not affected this way, the phonon strengths A1 and A2 derived from the measured spectra depend uncontrollably on pressure. This influence is eliminated by determining the ratio A1/A2 of both components shown in Figure 20. Based on the description of the 270/320 cm^{-1} phonon doublet above, Figure 21 indicates a strong site exchange of polar C atoms within the phase transition around 32 GPa.

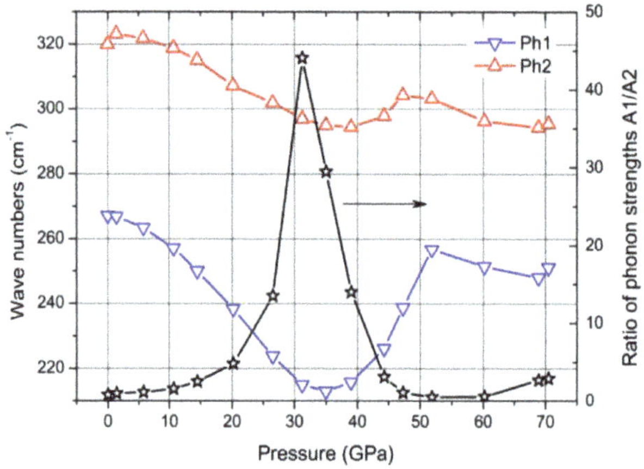

Figure 21. Wave numbers of the Raman-active 270/320 cm^{-1} doublet and ratio of the phonon strengths of both components.

7. Discussion

The electronic structure of boron carbide is dominated by $B_{11}C$ icosahedra. In comparison, the influence of the 3-atom chain is of minor importance. This is obvious in the phase transition near the chemical composition $B_{\sim 8}C$, which is related to the concentration of $B_{11}C$ icosahedra. The fact that this transition occurs at about 25% $B_{11}C$ instead of 50% shows the particular influence of the polar C atom that is missing in the frequently used hypothetical structure model $B_{12}CBC$. Whether the distribution of one C atom on six possible polar sites has any effect within this composition-dependent phase transition remains open.

However, the distribution of polar C atoms is doubtlessly important in $B_{4.3}C$ with ~100% $B_{11}C$ icosahedra. The phase transitions at 712 K and 33.2 GPa are directly related to the distribution of the polar C atoms. At ambient conditions, boron carbide is opaque. Its optical absorption in the visible range of the optical spectrum is high due to optical transitions related to gap states shown in Figure 4. Hushur et al. [43,55] showed that high pressure changes the opacity drastically to transparency, indicating a highly ordered structure. However, this is metastable only, as it retransforms into the disordered state, reverting to normal conditions, albeit with considerable relaxation time. Accordingly, the well-known distorted structure is energetically favored at ambient conditions. Applying this knowledge to the hypothetical structure $B_{12}CBC$ possibly explains why this highly ordered structure is unstable compared with the distorted structure of natural boron carbide.

This leads to the conclusion that Ektarawong's RNG model, assuming a random distribution of polar C atoms over all the polar sites at ambient conditions, is close to reality. The phase transitions at 712 K and 33.2 GPa both transform this distorted structure of boron carbide at ambient conditions (statistical distribution of 1 C atom per icosahedron to 6 polar sites) into a structure of higher order. This means the occupation of preferred polar sites; however, the actual distribution remains open. More detailed characterization of the phonon spectra might help.

In this study, the phonon spectra of boron carbides with natural isotope distribution and those highly enriched with ^{11}B isotopes have been analyzed. Thus, isotopes could have a certain effect (for details, see [64]). The force constants of different isotopes are the same. Hence, the shift of the icosahedral phonon frequencies in $B_{4.3}C$ boron carbides, which are enriched with ^{12}C and ^{13}C, respectively (see spectra in [40,65]), yields the relation of the reduced masses involved.

Funding: This research received no external funding.

Institutional Review Board Statement: Not applicable.

Informed Consent Statement: Not applicable.

Data Availability Statement: Data are available on request.

Acknowledgments: The author thanks all coauthors of joint papers cited in this review for their contribution particularly Murli Manghnani, University of Hawaii, Honolulu, USA, Karol Flachbarth, Institute of Experimental Physics, Slovak Academy of Sciences, Košice, Slovakia and Volodimir Filipov, I.N. Frantsevich Institute for Problems of Materials Science, Kyiv, Ukraine, for their productive cooperation during the last years. Moreover, thanks are due to the Journal of Solid-State Science (Elsevier Masson SAS) and the Journal of Physic: Condensed Matter (IOP Publishing) for the copyright permission of several figures reused in this review.

Conflicts of Interest: The author declares no conflict of interest.

References

1. Lipp, A. Borkarbid–Herstellung -Eigenschaften–Verwendung. *Tech. Rundsch.* **1965**, 57 (Heft 14) 5-9; 57 (Heft28) 19-23; 57 (Heft 33) 5-7; 1966, 58 (Heft 7) 3-11.
2. Domnich, V.; Reynaud, S.; Haber, R.A.; Chhowalla, M.J. Boron Carbide: Structure, Properties, and Stability under Stress. *Am. Ceram. Soc.* **2011**, *94*, 3605. [CrossRef]
3. Werheit, H. On Microstructure and Electronic Properties of Boron Carbide. *Ceram. Eng. Sci. Proc.* **2014**, *35*, 87–102.
4. Werheit, H. Structural Defects: Essential elements of icosahedral boron-rich solids. *Sustain. Ind. Process. Summit* **2015**, *8*, 159.
5. Werheit, H. Thermoelectric Properties of Boron-Rich Solids and their Possibilities of Technical Application. In Proceedings of the 2006 25th International Conference on Thermoelectrics, Vienna, Austria, 6–10 August 2006; p. 159.
6. Pristáš, G.; Werheit, H.; Gabáni, S.; Shalamberidze, S.; Flachbart, K. Low temperature specific heat anomaly with boson peak in isotope-enriched boron carbides $B_{4.3}C$–$B_{10}C$. *Solid State Sci.* **2020**, *101*, 106140. [CrossRef]
7. Qu, Z.; Yu, C.; Wei, Y.; Su, X.; Du, A.J. Thermal conductivity of boron carbide under fast neutron irradiation. *Adv. Ceram.* **2022**, *11*, 482.
8. Schwetz, K.A.; Karduck, P.J. Investigations in the Boron-Carbon System with the Aid of Electron Probe Microanalysis. *Less Common. Met.* **1991**, *175*, 1. [CrossRef]
9. Schwetz, K.A.; Karduck, P. Investigations in the Boron-Carbon System with the Aid of Electron Probe Microanalysis. *AIP Conf. Proc.* **1990**, *231*, 405.
10. Werheit, H.; Leithe-Jasper, A.; Tanaka, T.; Rotter, H.W.; Schwetz, K.A. Some properties of single-crystal boron carbide. *J. Solid State Chem.* **2004**, *177*, 575. [CrossRef]
11. Werheit, H.; Schwetz, K.A. Comment on "The Role of Free Carbon in the Transport and Magnetic Properties of Boron Carbide by A. K. Bandyopadhyay, F. Beuneu, L. Zuppiroliand, M. Beauvy". *J. Phys. Chem. Solids.* **1985**, *46*, 1331. [CrossRef]
12. Bandyopadhyay, A.K.; Beuneu, F.; Zuppiroli, L.; Beauvy, M.J. The Role of Free Carbon in the Transport and Magnetic Properties of Boron Carbide. *Phys. Chem. Solids* **1984**, *45*, 207. [CrossRef]
13. Werheit, H.; Manghnani, M.H.; Hushur, A. Phonon peculiarities at the high-pressure phase transition of $B_{4.3}C$ boron carbide. *Solid State Sci.* **2019**, *97*, 105978. [CrossRef]
14. Werheit, H.; Flachbart, K.; Pristáš, G.; Lotnyk, D.; Filipov, V.; Kuhlmann, U.; Shitsevalova, N.; Lundström, T. Influence of dopants, particularly carbon, on β-rhombohedral boron. *Semicond. Sci. Technol.* **2017**, *32*, 095015. [CrossRef]
15. Werheit, H.; Kuhlmann, U. Isotopic phonon effects in β-rhombohedral boron—non-statistical isotope distribution. *J. Phys. Condens. Matter* **2012**, *24*, 305401. [CrossRef]
16. Werheit, H. Boron carbide: Consistency of components, lattice parameters, fine structure and chemical composition makes the complex structure reasonable. *Solid State Sci.* **2016**, *60*, 45. [CrossRef]
17. Werheit, H. Assignment of the electronic transitions in $B_{4.3}C$ boron carbide implies a specifically distorted crystal structure. *Solid State Sci.* **2018**, *86*, 38. [CrossRef]
18. Werheit, H. Systematic error in conventionally measured Raman spectra of boron carbide—A general issue in solid state Raman spectroscopy. *Rev. Sci. Instrum.* **2019**, *90*, 043114. [CrossRef]
19. Werheit, H. Comment on "Understanding first-order Raman spectra of boron carbides across the homogeneity range". *Phys. Rev. Mater.* **2022**, *6*, 016601. [CrossRef]
20. Werheit, H.; Filipov, V.; Kuhlmann, U.; Schwarz, U.; Armbrüster, M.; Leithe-Jasper, A.; Tanaka, T.; Higashi, I.; Lundström, T.; Gurin, V.N.; et al. Raman effect in icosahedral boron-rich solids. *Sci. Technol. Adv. Mater.* **2010**, *11*, 023001. [CrossRef]
21. Beckel, C.L.; Youssaf, M.; Fuka, M.Z.; Raja, S.Y.; Lu, N. Lattice vibrations of the icosahedral solid α-boron. *Phys. Rev. B* **1991**, *44*, 2535. [CrossRef]
22. Werheit, H.; Shalamberidze, S.J. Advanced microstructure of boron carbide. *Phys. Condens. Matter* **2012**, *24*, 385406. [CrossRef]
23. Korotaev, P.; Pokatashkin, P.; Yanilkin, A. A Study of Mode III Fracture Toughness in Young and Mature Concrete with Fly Ash Additive. *Comput. Mater. Sci.* **2016**, *12*, 106. [CrossRef]

24. Schwetz, K.A.; Hassler, J.J. A wet chemical method for the determination of free carbon in boron carbide, silicon carbide and mixtures thereof. *Less Common. Met.* **1986**, *117*, 7. [CrossRef]
25. Bylander, D.M.; Kleinman, L. Structure of $B_{13}C_2$. *Phys. Rev. B* **1991**, *43*, 1487. [CrossRef] [PubMed]
26. Kleinman, L. Ab initio calculations of boron and ist carbides. *AIP Conf. Proc.* **1991**, *231*, 13.
27. Werheit, H. On Microstructure and Electronic Properties of Boron Carbide. In Proceedings of the 38th International Conference Advanced Ceramics and Composites, Daytona Beach, FL, USA, 26–31 January 2014.
28. Werheit, H.; Gerlach, G.J. Dynamical conductivity of boron carbide: Heavily damped plasma vibrations. *Phys. Condens. Matter* **2014**, *26*, 425801. [CrossRef]
29. Rasim, K.; Ramlau, R.; Leithe-Jasper, A.; Mori, T.; Burkhardt, U.; Borrmann, H.; Schnelle, W.; Carbogno, C.; Scheffler, M.; Grin, Y. Local Atomic Arrangements and Band Structure of Boron Carbide. *Angew. Chem.* **2018**, *130*, 6238. [CrossRef]
30. Chuvashova, I.; Gasharova, B.; Mathis, Y.-L.; Dubrovinsky, L.; Dubrovinskaia, N.Z. Structural Stability of Boron Carbide under Pressure Proven by Spectroscopic Studies up to 73 GPa. *Anorg. Allg. Chem.* **2017**, *643*, 1357. [CrossRef]
31. Werheit, H. Comment on Structural Stability of Boron Carbide under Pressure Proven by Spectroscopic Studies up to 73 GPa. Z. *Anorg. Allg. Chem.* **2018**, *644*, 353. [CrossRef]
32. Roma, G.; Gillet, K.; Jay, A.; Vast, N.; Gutierrez, G. Understanding first-order Raman spectra of boron carbides across the homogeneity range. *Phys. Rev. Mater.* **2021**, *5*, 063601. [CrossRef]
33. Sahu, T.; Bhattacharyya, A.; Gandi, A.N. Raman spectra characterization of boron carbide using first-principles calculations. *Phys. B* **2022**, *633*, 413738. [CrossRef]
34. Werheit, H. Comment on "Raman spectra characterization of boron carbide using first-principles calculations". *Phys. B* **2023**, *651*, 414604. [CrossRef]
35. Hushur, A.; Manghnani, M.H.; Werheit, H.; Dera, P.; Williams, Q.J. High-pressure phase transition makes $B4_{.3}C$ boron carbide a wide-gap semiconductor. *Phys. Condens. Matter* **2016**, *28*, 045403. [CrossRef] [PubMed]
36. Werheit, H.; Manghnani, M.H.; Kuhlmann, U.; Hushur, A.; Shalamberidze, S. Mode Grüneisen parameters of boron carbide. *Solid State Sci.* **2017**, *72*, 80. [CrossRef]
37. Werheit, H.; Herstell, B.; Winkelbauer, W.; Pristáš, G.; Gabáni, S.; Flachbart, K.; Shalamberidze, S. Electrical conductivity of boron carbide from ~5 to ~2100 K in the whole homogeneity range. *Solid State Sci.* **2022**, *132*, 106987. [CrossRef]
38. Wood, C. Transport properties of boron carbide. *AIP Conf. Proc.* **1986**, *140*, 206.
39. Wood, C. The transport properties of boron carbide prepared by different techniques. In Proceedings of the 9th International Symposium on Boron, Borides and Related Compounds, Duisburg, Germany, 21–25 September 1987; Werheit, H., Ed.; University of Duisburg: Duisburg, Germany, 1987; p. 213.
40. Aselage, T.L.; Emin, D.; McCready, S.S. Conductivities and Seebeck coefficients of boron carbides: Softening bipolaron hopping. *Phys. Rev. B* **2001**, *64*, 054302. [CrossRef]
41. Amulele, G.; Manghnani, M.H.; Werheit, H.; Lanati, A.W.; Clark, S.M. Electrical conductivity of B~4.3C boron carbide up to 9 GPa and 1273 K. *Solid State Sci.* **2023**, *140*, 107185. [CrossRef]
42. Werheit, H. On excitons and other gap states in boron carbide. *J. Phys. Condens. Matter* **2006**, *18*, 10655. [CrossRef]
43. Feng, Y.; Seidler, G.T.; Cross, J.O.; Macrander, A.T.; Rehr, J.J. Role of inversion symmetry and multipole effects in nonresonant x-ray Raman scattering from icosahedral B_4C. *Phys. Rev. B* **2004**, *69*, 125402. [CrossRef]
44. Schmechel, R.; Werheit, H. Structural Defects of Some Icosahedral Boron-Rich Solids and Their Correlation with the Electronic Properties. *Solid State Chem.* **2000**, *154*, 61. [CrossRef]
45. Armstrong, D.R.; Balland, J.; Perkins, P.G.; Will, G.; Kirfel, A. The Nature of the Chemical Bonding in Boron Carbide. IV. Electronic Band Structure of Boron Carbide, $B_{13}C_2$, and Three Models of the Structure $B_{12}C_3$. *Acta Crystallogr. B* **1983**, *39*, 324. [CrossRef]
46. Bullett, D.W. Structure and bonding in crystalline boron and $B_{12}C_3$. *J. Phys. C Solid State Phys.* **1982**, *15*, 415. [CrossRef]
47. Switendick, A.C. *The Electronic Structure of Crystalline Boron Carbide I: B_{12} Icosahedra and C–B–C Chains in The Physics and Chemistry of Carbides, Nitrides and Boride*; Freer, R., Ed.; Kluwer Academic: Dordrecht, The Netherlands, 1990; p. 525.
48. Bylander, D.M.; Kleinman, L.; Lee, S. Self-consistent calculations of the energy bands and bonding properties of $B_{12}C_3$. *Phys. Rev. B* **1990**, *42*, 1394. [CrossRef] [PubMed]
49. Shirai, K.; Dekura, H.; Masago, A. Superconductivity research on boron solids and an efficient doping method. *J. Phys. Conf. Ser.* **2009**, *176*, 012001. [CrossRef]
50. Calandra, M.; Vast, N.; Mauri, F. Superconductivity from doping boron icosahedra. *Phys. Rev. B* **2004**, *69*, 224505. [CrossRef]
51. Vast, N.; Sjakste, J.; Betranhandy, E.J. Boron carbides from first principles. *Phys. Conf. Ser.* **2009**, *176*, 012002. [CrossRef]
52. Werheit, H.; Au, T.; Schmechel, R.; Shalamberidze, S.O.; Kalandadze, G.I.; Eristavi, A.M. IR-Active Phonons and Structure Elements of Isotope-Enriched Boron Carbide. *J. Solid State Chem.* **2000**, *154*, 79. [CrossRef]
53. Werheit, H.; Rotter, H.W.; Meyer, F.D.; Hillebrecht, H.; Shalamberidze, S.O.; Abzianidze, T.G.; Esadze, G.G. FT-Raman spectra of isotope-enriched boron carbide. *J. Solid State Chem.* **2004**, *177*, 569. [CrossRef]
54. Shirai, K. Elastic properties and the mechanical stability of icosahedral boron crystals. *Phys. Rev. B* **1997**, *55*, 12235. [CrossRef]
55. Shirai, K.; Emura, S. Lattice Vibrations of Boron Carbide. *J. Solid State Chem.* **1997**, *133*, 93. [CrossRef]
56. Lazzari, R.; Vast, N.; Besson, J.M.; Baroni, S.; Dal Corso, A. Atomic Structure and Vibrational Properties of Icosahedral B_4C Boron Carbide. *Phys. Rev. Lett.* **1999**, *83*, 3230; Erratum in *Phys. Rev. Lett.* **2000**, *85*, 4194. [CrossRef]

57. Kuhlmann, U. Zusammenhänge zwischen den Phononenspektren borreicher Festkörper mit Ikosaederdtruktur und ihren Strukturellen und Elektronischen Eigenschaften. Ph.D. Thesis, University of Duisburg, Duisburg, Germany, 1994.
58. Werheit, H.; Hoffmann, S.; Gerlach, G.; Leithe-Jasper, A.; Tanaka, T. Composite, Quasi-crystals and Nanomaterials. In *Phase Transition in B4.3C Boron Carbide at Moderate Temperature 2015 Sustainable Industrial Processing Summit*; Kongoli, F., Pech-Canul, M., Kalemptas, A., Werheit, H., Eds.; FLOGEN: Antalya, Turkey, 2015; Volume 8, p. 183.
59. Yao, S.; Huhn, W.P.; Widom, M. Phase transitions of boron carbide: Pair interaction model of high carbon limit. *Solid State Sci.* **2015**, *47*, 21. [CrossRef]
60. Ektarawong, A.; Simak, S.I.; Hultman, L.; Birch, J.; Alling, B. First-principles study of configurational disorder in B_4C using a superatom-special quasirandom structure method. *Phys. Rev. B* **2014**, *90*, 024204. [CrossRef]
61. Dekura, H.; Shirai, K.; Yanase, A. Metallicity of boron carbides at high pressure. *J. Phys. Conf. Ser.* **2010**, *215*, 012117. [CrossRef]
62. Werheit, H. Van Hove singularities of some icosahedral boron-rich solids by differential reflectivity spectra. *Solid State Sci.* **2015**, *47*, 13. [CrossRef]
63. Werheit, H.; Janowitz, C.; Schmechel, R.; Tanaka, T.; Ishizawa, Y. Interband Critical Points of Some Icosahedral Boron-Rich Solids. *J. Solid State Chem.* **1997**, *133*, 132. [CrossRef]
64. Werheit, H.; Kuhlmann, U.; Rotter, H.W.; Shalamberidze, S.O. Isotopic effects on the phonon modes in boron carbide. *J. Phys. Condens. Matter* **2010**, *22*, 395401. [CrossRef] [PubMed]
65. Werheit, H. Isotopic phonon effects in boron-rich solids. *Solid State Sci.* **2012**, *14*, 1559. [CrossRef]

Disclaimer/Publisher's Note: The statements, opinions and data contained in all publications are solely those of the individual author(s) and contributor(s) and not of MDPI and/or the editor(s). MDPI and/or the editor(s) disclaim responsibility for any injury to people or property resulting from any ideas, methods, instructions or products referred to in the content.

materials
MDPI

Review

Composites and Materials Prepared from Boron Cluster Anions and Carboranes

Varvara V. Avdeeva [1,*], Svetlana E. Nikiforova [1], Elena A. Malinina [1], Igor B. Sivaev [2,3] and Nikolay T. Kuznetsov [1]

1 Kurnakov Institute of General and Inorganic Chemistry, Russian Academy of Sciences, 31 Leninskii Av., Moscow 119991, Russia; korolencko0110@yandex.ru (S.E.N.); malinina@igic.ras.ru (E.A.M.); ntkuz@igic.ras.ru (N.T.K.)
2 Nesmeyanov Institute of Organoelement Compounds, Russian Academy of Sciences, 28 Vavilov Str., Moscow 119991, Russia; sivaev@ineos.ac.ru
3 Basic Department of Chemistry of Innovative Materials and Technologies, Plekhanov Russian University of Economics, 36 Stremyannyi Line, Moscow 117997, Russia
* Correspondence: avdeeva.varvara@mail.ru

Abstract: Here, we present composites and materials that can be prepared starting with boron hydride cluster compounds (decaborane, decahydro-*closo*-decaborate and dodecahydro-*closo*-dodecaborate anions and carboranes). Recent examples of their utilization as boron protective coatings including using them to synthesize boron carbide, boron nitride, metal borides, metal-containing composites, and neutron shielding materials are discussed. The data are generalized demonstrate the versatile application of materials based on boron cluster anions and carboranes in various fields.

Keywords: boron clusters; carborane; composites; borides

1. Introduction

The driving force behind the intensive development of the chemistry of boron hydride cluster compounds in the 1950s was their use as high-energy materials [1–3]. This topic is still discussed, although it attracts less attention [4–14]. Currently, the most attention is paid to the use of boron cluster compounds in medicine [15–27]. As for the use of boron clusters in materials science, complex molecular and supramolecular structures, such as nanocars and nanotrains [28–31], molecular machines, and switches [32–36], MOFs [37–46], etc., attract the most attention. However, there is another less spectacular but no less important direction of research on the use of boron cluster compounds in the science of materials. This direction of research consists of the thermal decomposition of various boron clusters with the formation of boron-containing coatings and ceramic materials, which is somewhat reminiscent of their use as high-energy materials, since in this case a complex molecular structure is also converted into simple molecules.

In this contribution to the field, we present an attempt to consider the use of boron clusters to obtain various boron-containing materials, including boron-containing coatings and films, nanostructured boron carbide and metal borides, and others. The most readily available decaborane(14) $B_{10}H_{14}$, *ortho*- and *meta*-carboranes 1,2-$C_2B_{10}H_{12}$ and 1,2-$C_2B_{10}H_{12}$, decahydro-*closo*-decaborate $[B_{10}H_{10}]^{2-}$ and octadecahydro-*conjucto*-eicosaborate [*trans*-$B_{20}H_{18}]^{2-}$ anions (Figure 1) are considered as the boron clusters. Decaborane(14) can be readily prepared in a two-step procedure from sodium tetrahydroborate $NaBH_4$ [47]. The well-known *ortho*-carborane is obtained by introducing the acetylene molecule into the open boron cage of decaborane(14), and its thermal isomerization leads to *meta*-carborane [48]. The decahydro-*closo*-decaborate anion [49–51] is formed by heating decaborane(14) in the presence of triethylamine, and its mild oxidation results in the octadecahydro-*conjucto*-eicosaborate anion [52,53].

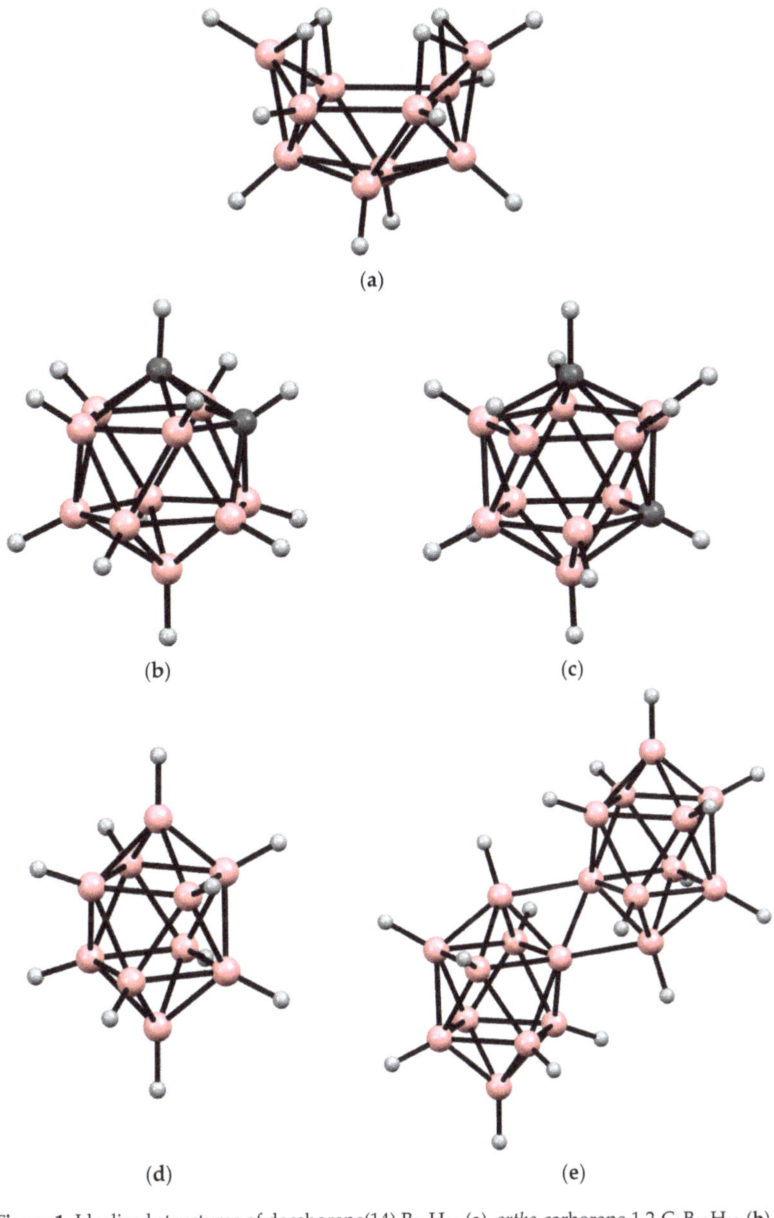

Figure 1. Idealized structures of decaborane(14) $B_{10}H_{14}$ (**a**), *ortho*-carborane 1,2-$C_2B_{10}H_{12}$ (**b**), *meta*-carborane 1,7-$C_2B_{10}H_{12}$ (**c**), decahydro-*closo*-decaborate anion $[B_{10}H_{10}]^{2-}$ (**d**), and octadecahydro-*conjucto*-eicosoborate anion [*trans*-$B_{20}H_{18}]^{2-}$ (**e**).

2. Decaborane as Boron Source for Boron-Containing Materials

Metal Organic Chemical Vapor Deposition (MOCVD) is widely used for creating high-purity crystalline semiconducting thin films and micro/nano structures for microelectronics [54–57]. Therefore, it is not surprising that volatile boron hydrides such as decaborane have been proposed for the preparation of various boron-containing coatings. At first glance, it may seem that for these purposes it is more convenient to use other more available and cheap volatile boron compounds, such as BCl_3, BF_3 or diborane. However, one drawback of MOCVD is the aggressive, toxic and explosive nature of the precursor gases, which makes them difficult to use in small research laboratories. This fully applies to both aggressive and corrosive boron halides and highly toxic and flammable diborane. Therefore, decaborane $B_{10}H_{14}$, despite its higher cost and toxicity [58,59], is in many respects a more convenient source of boron for these purposes. Furthermore, decaborane has advantages as a source material for boron coating because high-purity decaborane is easy to obtain with a sublimation purification process.

Amorphous boron films of 0.1–1.5 μm thickness have been prepared on sapphire, silicon, and tantalum as substrates by the pyrolysis of decaborane in the molecular flow region ($\leq 10^{-4}$ torr) and in a temperature range of 350–1200 °C. It is found that the deposition rate of the boron films is proportional to the decaborane partial pressure and the substrate temperature. The electrical conductivities vary from 3×10^{-5} S cm^{-1} at 77 K to 30 S cm^{-1} at 1000 K, and the activation energy is 1.07 eV in the intrinsic temperature range (700–1000 K). The maximum value of thermoelectric power is about 420 μV deg^{-1} at 700 K, and its polarity is positive between 500 and 1000 K [60]. The preparation of polycrystalline α-rhombohedral boron films by pyrolysis of decaborane has also been reported [61].

Boron coating also can be obtained by the plasma-assisted enhanced chemical vapor deposition (PECVD) of decaborane [62,63]. In particular, decaborane has been proposed as a boron source for the boronization of JT-60U Tokamak to reduce the influx of impurities during plasma discharge [64]. The deposition of boron films on polished *p*-type Si(111) surface by synchrotron-radiation-induced chemical vapor deposition (SR-CVD) of decaborane was reported [65].

When using ammonia or dinitrogen as additives, the chemical vapor deposition of decaborane can be used to obtain thin films of boron nitride [66], nanosheets [67], and nanotubes [68], competing with borazine. In particular, boron nitride nanotubes (BNNTs) grown at 1200–1300 °C from decaborane were double- and multiwalled, with the double-walled nanotubes having ~2 nm inner diameters and the multiwalled nanotubes (~10 walls) having ~4–5 nm inner diameters and ~12–14 nm outer diameters. The nanotubes grown at 1300 °C were longer, averaging ~0.6 μm, whereas those grown at 1200 °C had average lengths of ~0.2 μm [68].

The pyrolysis of decaborane can also be used to prepare boron nanoparticles [69] and microcrystals [70,71]. α-Tetragonal boron crystals were obtained at a pressure of 8–9 GPa and temperatures in the range 1100–1600 °C, while β-rhombohedral boron crystals grow at 3 GPa and 1200 °C [70]. The α-tetragonal boron crystals synthesized demonstrate semiconducting properties of conductivity with the energy gap $E_g \approx 1.5$ eV [71].

The chemical vapor deposition of decaborane was also used to prepare various metal boride thin films, including nickel [72], strontium [73], gadolinium [74], neodymium [75], and ytterbium [76,77].

A synthetic route to metal borides TiB_2, $ZrBz_2$, HfB_2, NbB_2, and TaB_2 by heating the decaborane-pimelonitrilium polymer [-6-$B_{10}H_{12}$-(NC(CH_2)$_5$CN)]$_n$- and the corresponding finely dispersed metal oxides above 1400 °C was proposed. The metal boride powders were found to be highly crystalline, with grain sizes dependent on processing temperatures [78].

Various boron-carbide-containing materials were prepared using various decaborane-based single-molecular precursors, such as [μ-6,6′-(CH_2)$_6$-($B_{10}H_{13}$)$_2$] [79,80], [μ-6,6′-(1′,5′-cyclooctyl)-($B_{10}H_{13}$)$_2$] [81], [μ-6,6′-(2′,5′-norbornenyl)-($B_{10}H_{13}$)$_2$] [81], or polymers, including [-6-$B_{10}H_{12}$-Ph_2POPPh_2]$_n$- [82], [-6-$B_{10}H_{12}$-(CH_2)$_6$]$_n$- [83,84], [-6-$B_{10}H_{12}$-(2′,5′-norbornenyl)]$_n$- [81,83–89], [-6-$B_{10}H_{12}$-(1′,5′-cyclooctenyl)]$_n$- [81,83], and [-6-$B_{10}H_{12}$-(1′,4′-cyclooctenyl)]$_n$- [81,83].

In particular, [μ-6,6′-(CH_2)$_6$-($B_{10}H_{13}$)$_2$] (Figure 2) appears to be an ideal precursor for the synthesis of boron carbide nanofibers (Figure 3) using the templating technique: (i) it is readily synthesized in large amounts using the Ti-catalyzed reaction [90]; (ii) it contains no other ceramic-forming elements and has a desirable boron-to-carbon ratio, thus yielding boron-rich boron carbide compositions upon pyrolysis; (iii) it is stable as a liquid, allowing it to be absorbed into the membrane without decomposition; and (iv) upon pyrolysis, it undergoes a cross-linking reaction at relatively low temperatures (220 °C), which slows the loss of material by volatilization, thereby generating high ceramic and chemical yields [79].

Figure 2. Solid-state structure of [μ-6,6′-$(CH_2)_6$-$(B_{10}H_{13})_2$]. Hydrogen atoms of organic substituents are omitted for clarity.

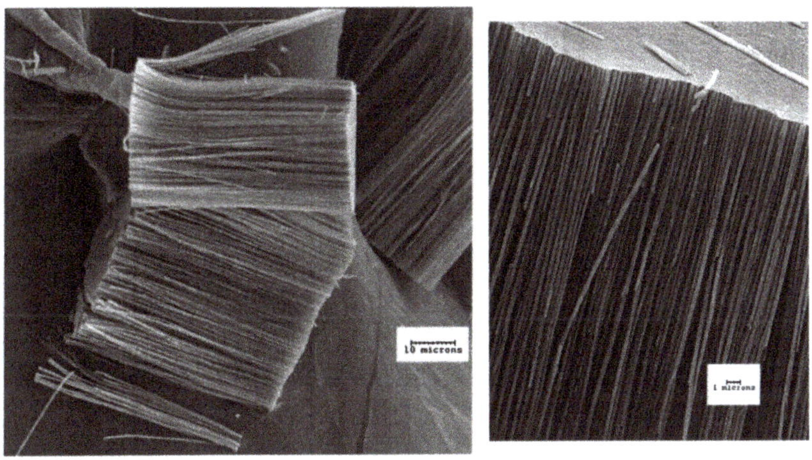

Figure 3. SEM images of aligned boron carbide nanofibers obtained upon pyrolysis of [μ-6,6′-$(CH_2)_6$-$(B_{10}H_{13})_2$] at 1025 °C. Reprinted with permission from Ref. [79]. Copyright (2000) the American Chemical Society.

The bis(decaboranyl)-hexane precursor [μ-6,6′-$(CH_2)_6$-$(B_{10}H_{13})_2$] can also be used for the preparation of ordered mesoporous boron carbide materials with high specific surface areas up to 778 m^2/g and hexagonal pore arrangement symmetries [80].

It should be noted that an alternative possibility of using pentaborane B_5H_9 instead of decaborane to obtain boron carbide compositions was previously considered [91–93]; however, after the destruction of pentaborane stocks stored since the 1960s [94], this aim was abandoned.

The B_4C/BN-containing ceramic materials can be prepared using the pyrolysis of polymeric Lewis base adducts of decaborane [-6-$B_{10}H_{12}$-(diamine)]$_n$- (diamine is ethylenediamine, 1,1-dimethylethylenediamine, 1,1,2,2-tetramethylethylenediamine) [95,96]. In particular, the [-6-$B_{10}H_{12}$-(ethylenediamine)]$_n$-polymer fibers upon pyrolysis at 1000 °C in an argon atmosphere retain their shape and give black ceramic fibers with a diameter of 3 to 5 μm, which have a round shape, a smooth surface, and no obvious major flaws (Figure 4). Other [-6-$B_{10}H_{12}$-(diamine)]$_n$- polymers were capable of forming fibers. The polymers derived from 1,1,2,2-tetramethylethylenediamine and from the 85/15 1,1-dimethylethylenediamine/1,1,2,2-tetramethylethylenediamine mixture melt when heated (mp 246–250 °C and 222–225 °C, respectively) and may be suitable for melt-spinning [87].

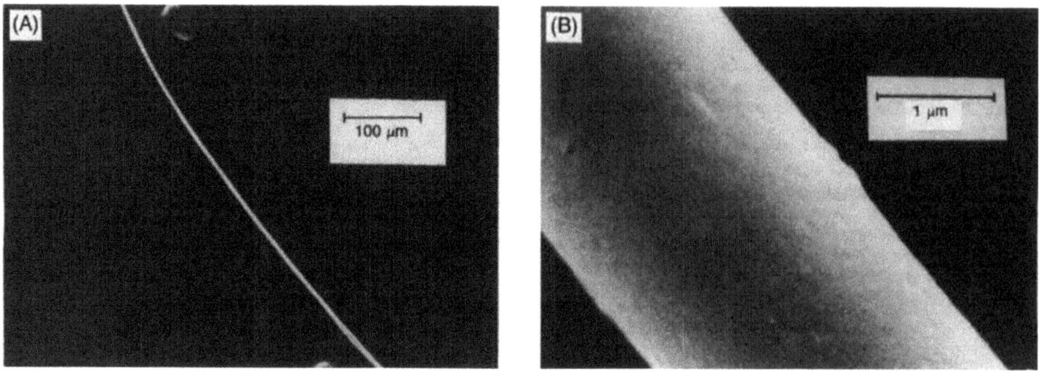

Figure 4. SEM images of ceramic fibers at different scale (**A**,**B**) derived from [-6-$B_{10}H_{12}$-(ethylenediamine)]$_n$- by pyrolysis at 1000 °C under argon. Reprinted with permission from Ref. [87]. Copyright (1988) the American Ceramic Society.

Low-crystalline boron nitride was prepared by the reaction of triammoniadecaborane $B_{10}H_{14}{\cdot}3NH_3$ and hydrazine or ammonia at 125 MPa and 650–700 °C. The prepared low-crystalline boron nitride passed into cubic boron nitride at 1200–1300 °C and 6.5 GPa in the presence of 20 mol.% AlNas a catalyst [97,98].

3. Carborane as a Boron Source for Boron-Containing Materials

Chemical vapor deposition methods are widely used for the manufacture of boron carbide films due to the better controlled deposition process and the high-quality boron carbide production. Mixtures of boron trichloride, methane, and hydrogen are usually used for the CVD of boron carbide films. Since chlorides are highly dangerous and the synthesis process requires high temperatures, the replacement of BCl_3 with organoboranes has become a trend in recent years. At first it seemed that small organoboron molecules such as trimethylboron and triethylboron could be a good alternative, but they proved to be overly reactive. Taking into account that *ortho*-carborane [$H_2C_2B_{10}H_{10}$] provides a suitable ratio of B and C from a single molecular source, it seems to be an attractive source for preparing boron carbide materials.

Semiconducting boron carbide represents a new class of materials with potential applications in neutron detection because ^{10}B has a high cross-section (approximately 3800 barns) for neutrons at lower energies (~25 meV), based on the $^{10}B(n,\alpha)^7Li$ neutron capture reaction [99–105]. This aroused great interest in the fabrication of boron carbide films using the PECVD of *ortho*-carborane, and the effect of the process parameters, such as temperature and total pressure, on the composition, microstructure, morphology, and properties of the boron carbide films obtained were studied [106–114]. In particular, the boron carbide film prepared at low temperatures and pressures (T_{dep} = 900 °C and P_{tot} = 100 Pa) showed a comparatively flat morphology, whereas the boron carbide films prepared at low temperature and high pressure (T_{dep} = 900 °C and P_{tot} = 50,000 Pa) appeared as round bulges. The boron carbide films prepared at a high temperature and relatively low pressure (T_{dep} = 1100 °C and P_{tot} = 5000 Pa) exhibited a cauliflower-like surface, while the films prepared at high temperature and high pressure (T_{dep} = 1200 °C and P_{tot} = 50,000 Pa) exhibited a uniform granular surface (Figure 5) [112].

Semiconducting boron carbide films can be also prepared through the PECVD of *meta*-carborane, which differs from *ortho*-carborane only in the arrangement of carbon atoms in the icosahedral cage [115–117]. It was found that *meta*-carborane and *ortho*-carborane form self-doped *n*-type and *p*-type boron-carbides, respectively [115,116].

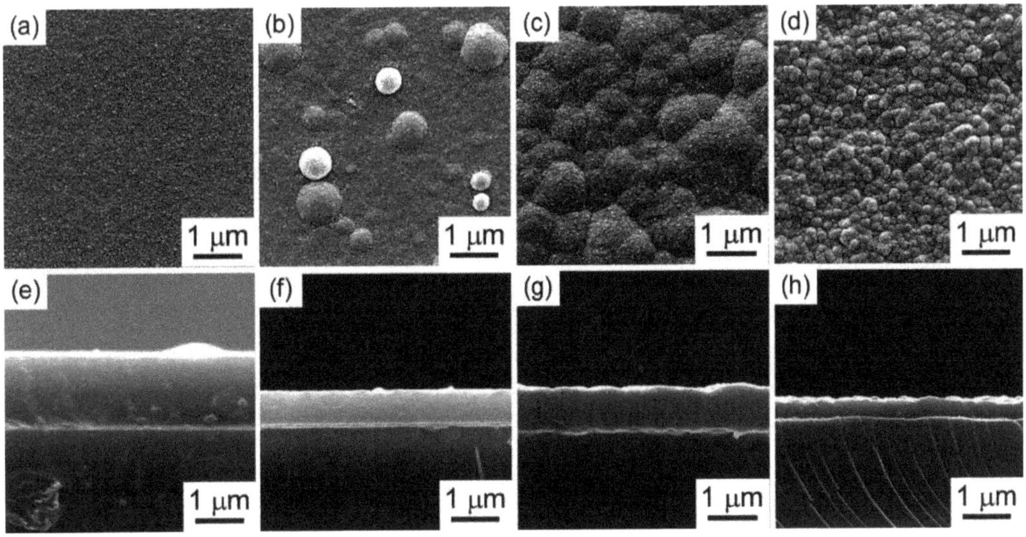

Figure 5. Surface and cross-section SEM images of boron carbide film prepared at T_{dep} = 900 °C, P_{tot} = 100 Pa (**a**,**e**), T_{dep} = 900 °C, P_{tot} = 50,000 Pa (**b**,**f**), T_{dep} = 1100 °C, P_{tot} = 5000 Pa (**c**,**g**), T_{dep} = 1200 °C, P_{tot} = 50,000 Pa (**d**,**h**). Reprinted from Ref. [112].

It was shown that neutron detectors and neutron voltaic devices, based on semiconducting boron carbides, contrary to most other electrical devices, may improve with some radiation exposure and are robust against radiation-induced device degradation and failure [104]. The main causes for the poor neutron detection device performance are the insufficiently thick depletion region of the device, the need for a thicker device to come closer to neutron opacity, and the need for better charge collection while maintaining low reverse bias leakage currents [118]. It was found that the semiconducting boron carbide prepared by PECVD of composites of *ortho*- and *meta*-carboranes and aromatic or heteroaromatic compounds [119–125] demonstrate improvements in both charge collection and reverse bias leakage currents, which is attributed to an increase in the hole carrier lifetimes.

The introduction of metallocenes Cp_2M (M = Ni, Co, Fe, Mn) together with *ortho*- or *meta*-carboranes during the PECVD process results in the corresponding transition metal doping of semiconducting boron carbide films [126–131].

Another important area of using *ortho*- and *meta*-carboranes to create protective boron carbide coatings is the boronization of tokamaks. The plasma-chemical deposition of a protective coating on the first wall of a fusion device using a chemically active gas (precursor) remains to date one of the primary ways to protect plasma against cooling impurities. This method has proved effective and does not require the use of additional and expensive equipment. The use of a low-toxic and nonexplosive carborane for boronization made this method of obtaining boron-carbon coatings to be a quick, widely available, and relatively cheap one. The coatings obtained were found to be highly resistant to chemical erosion—the erosion coefficients were (5–6) $\times 10^{-4}$ at/ion regardless of temperature. The electrical resistance of the coating was high, and, depending on the deposition conditions, varied in the range of 10^9–10^{11} Ω cm. The resistance of the coatings to the plasma impact was estimated using similar probes, which were examined after a certain number of working pulses. On all tokamaks, the coatings remained for several hundred pulses. The degradation of the coating correlates with the degradation of the plasma parameters [132–136].

The preparation of boron carbide through the pyrolysis of various carborane-containing polymers has been described [137–140]. The Ni-catalyzed polymerization of 1,2-bis(4-

chloro-phenyl)-*ortho*-carborane leads to poly(phenylene-*ortho*-carborane). It was found that the heating of the polymer at 1000–1200 °C resulted in the crystallization of boron carbide, according to the X-ray powder diffraction studies [139]. The Ni-catalyzed polymerization of 1,7-bis(4-chlorophenyl)-*meta*-carborane produces poly(phenylene-*meta*-carborane), which can be used as a novel boron carbide precursor [140]. Due to its high ceramic yield, it can be used to prepare boron carbide ceramics with different shapes [141]. In particular, poly(phenylene-*meta*-carborane) was used to prepare the boron carbide hollow microsphere via slurry-coating and a method derived from a previous study. The poly(phenylene-*meta*-carborane)/polyacrylonitrile slurry was prepared and coated on a polyoxymethylene ball substrate. After air cross-linking, the substrate decomposition and heat-treatment at 1100 °C in argon atmosphere, hollow boron carbide microspheres with diameter of approximate 1.34 mm, and average shell thickness of 30 μm were obtained (Figure 6) [141].

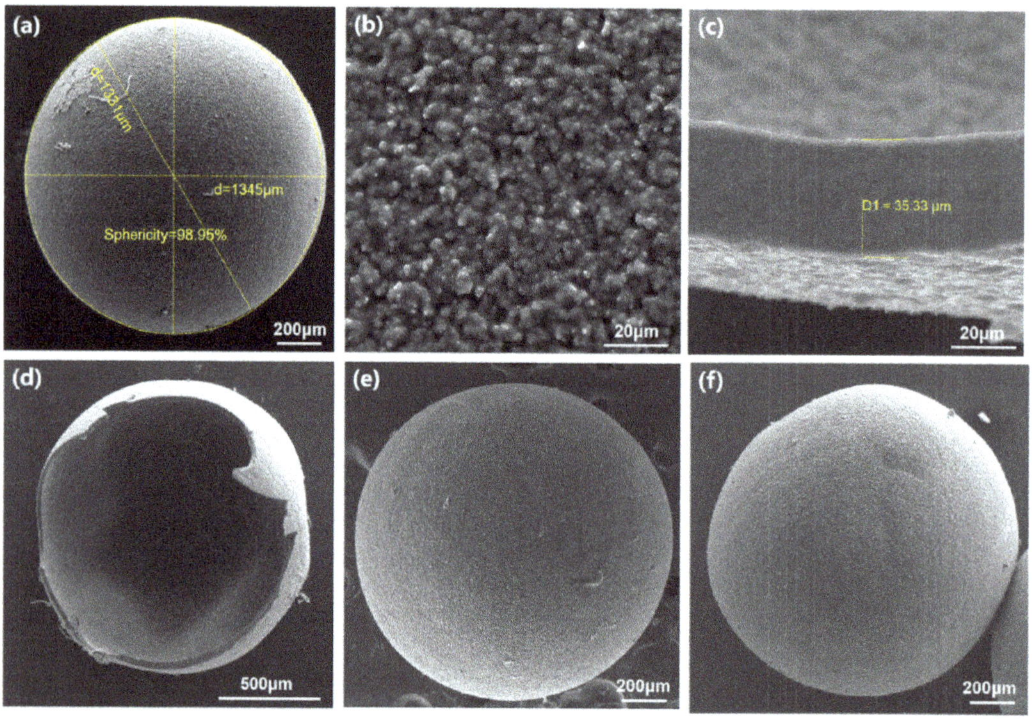

Figure 6. Hollow boron carbide microspheres prepared from poly(phenylene-*meta*-carborane)/polyacrylonitrile at different magnitude (**a**–**f**). Reprinted with permission from Ref. [141]. Copyright (2022) Elsevier.

The star-shaped pentagonal microcrystals of boron carbide with extremely low carbon content (~5%) were prepared through the thermobaric treatment of 1,7-bis(hydroxymethyl)-*meta*-carborane under high pressure of 7 GPa and temperature of 1370 K. The microcrystals exhibit a five-fold symmetry and grow in the shape of stars (Figure 7) [142,143]. The unusual shape of the pentagonal microcrystals makes them unique for developing novel micro-machines and semiconductor micro-devices [142].

Heating a mixture of *ortho*-carborane and adamantane (atomic ratio B:C = 5:95) at 8 GPa and 1700 °C results in the formation of boron-doped diamond microcrystals (2–2.5 at.% of boron), whereas only graphite was obtained from a mixture of adamantane and *ortho*-carborane at pressures lower than 7 GPa [144].

Figure 7. SEM images of star-shaped pentagonal boron carbide microcrystals prepared by thermobaric treatment of 1,7-bis(hydroxymethyl)-*meta*-carborane at 7 GPa and 1370 K. Reprinted from Ref. [142].

4. Boron Cluster Anions as Boron Source for Boron-Containing Materials

Coordination chemistry of transition metals with boron cluster anions is one of the most intensively studied fields of boron chemistry [145–147]. Research in this area is determined mainly by the fundamental components and concerns metal-boron cluster binding [148], positional isomerism [149,150], and secondary and interligand/inner-ligand interactions in complexes [151,152]. A series of new complex compounds that formed precursors and materials with desired properties were synthesized. Among them are precursors for the low-temperature synthesis of borides and related compounds [153], molecular switches based on a dimeric boron cluster [154], catalysts in the synthesis of organic compounds [155], complexes with luminescent properties [156], copper complexes as models for studying exchange processes and magnetic materials [157], as well as neutron-absorbing materials based on salts of boron cluster anions distributed in the silicate matrix.

Metal borides and related compounds provide ample opportunities for multivariate combination of metal–metal, metal–boron and boron–boron bonds in the resulting phases, thereby providing the possibility of directed changes in their physical, chemical and strength properties [157–162]. The most well-known methods of preparing metal borides include: (i) the reaction of metals and boron; (ii) the reduction of metal and boron from oxides when allowing to react with carbon or metals; (iii) the electrolytic reduction of metal and boron from their compounds; and (iv) the thermal dissociation of unstable compounds containing boron and metals. Actually, the processes used to prepare metal borides are often energy-consuming and time-consuming.

In the course of research carried out in our team, we have developed a method for obtaining binary borides during thermal reduction of transition metal compounds $[ML_x][An]$ (M = Co, Ni, An = $[B_{10}H_{10}]^{2-}$, $[B_{12}H_{12}]^{2-}$ or $[B_{20}H_{18}]^{2-}$) with ligands L that can be easily removed at elevated temperature (for example, L = H_2O, NH_3, DMF). In the compounds, organic ligands L are considered components, which play the role of organic fuel. Dimethylformamide is one of the most promising substances that can be used as a fuel [163]; its specific heat of combustion (29.652 MJ/kg) is much higher than, for example, that of urea (9.134 MJ/kg), which is often used in SCS processes. The energy capacity of the boron cluster anions themselves makes it possible to lower the boride synthesis temperature, which facilitates the process and reduces energy consumption.

First, we synthesized complexes $[Co(DMF)_6][B_{10}H_{10}]$ and $[Co(DMSO)_6][B_{10}H_{10}]$ (Figure 8), studied their thermooxidative properties in the temperature range 20–600 °C under argon [164], and determined the annealing temperature. When comparing the IR data of products of thermolysis performed at 600 °C, it was concluded that boride phases were prepared only for complex [Co(DMF)$_6$][$B_{10}H_{10}$]. The final products were X-ray amorphous that did not allow us to determine the exactly composition of the final products.

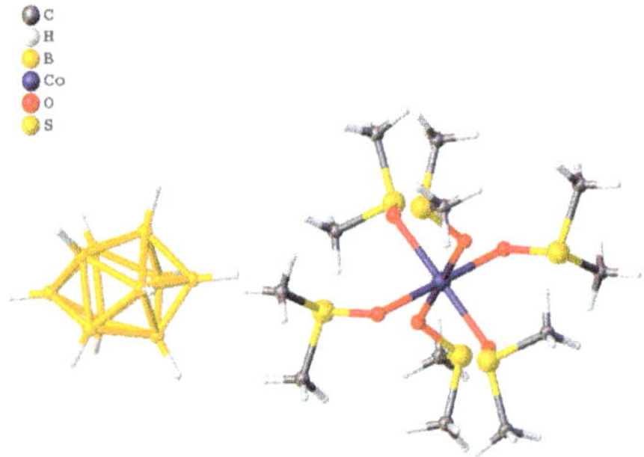

Figure 8. Structures of [Co(DMSO)$_6$][$B_{10}H_{10}$].

When annealing structurally related compounds [Co(DMF)$_6$][$B_{12}H_{12}$], [Co(DMF)$_6$][$B_{20}H_{18}$] (Figure 9) and [Co(DMF)$_6$][$B_{10}Cl_{10}$] in argon at 900 °C [165,166], we succeeded in detecting the CoB phase using X-ray powder diffraction [165]. It was found that for [Co(DMF)$_6$][$B_{12}H_{12}$], the phases of BN and CoB where prepared in the 1:1 ratio; for [Co(DMF)$_6$][$B_{20}H_{18}$], a higher CoB:BN ratio but low crystallinity were found; and for the cobalt(II) complex with the decachloro-*closo*-decaborate anion, only CoB was detected. The annealed samples were studied using IR spectroscopy and X-ray fluorescence (for the chloro-containing sample). The nanoparticular character of the decomposition products was shown using TEM.

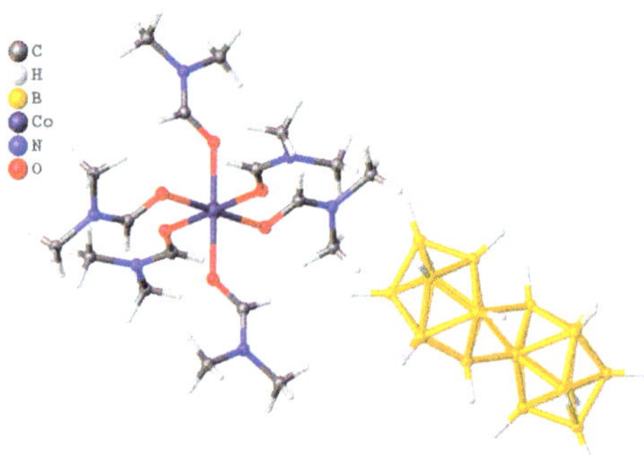

Figure 9. Structure of [Co(DMF)$_6$][$B_{20}H_{18}$].

Thermal reduction of complexes [CoL$_n$][$B_{10}H_{10}$] (L = H_2O, n = 6; N_2H_4, n = 3) with hydrazine and water molecules in argon at 650 and 900 °C [167,168] resulted in preparation of the dicobalt boride Co_2B phase as well as orthorhombic and cubic modifications of boron nitride BN. For the aquacomplex, oxide-boride phases were detected. The annealed samples were studied using IR spectroscopy and X-ray powder diffraction. In addition, the samples show different magnetochemical behavior: the oxide–boride phase demonstrated a significant ferromagnetic contribution to the total magnetization of the sample, while the nitride–boride phase had a diamagnetic contribution.

As for structurally related nickel complexes [NiL$_n$][$B_{10}H_{10}$] (L = DMF, H_2O, n = 6; L = N_2H_4, n = 3) [169], their thermal reduction was studied in the temperature range 20–800 °C in air and in argon. The phases of Ni_3C and $Ni_{1-x}C_x$ were detected using X-ray powder diffraction for annealed complex [Ni(DMF)$_6$][$B_{10}H_{10}$]; the obtained data indicates that boride-carbide phases were not detected.

Gadolinium tetraboride GdB_4 was found to form as an only-boride phase by heating a mixture of gadolinium hydride $GdH_{\sim 2}$ and gadolinium *closo*-decaborate $Gd_2[B_{10}H_{10}]_3$

as a boron source (the boron:metal ratio = 2) at 1400 °C under an argon atmosphere. In a similar way, cerium tetraboride CeB_4 was prepared from $CeH_{\sim 2}$ and $Ce_2[B_{10}H_{10}]_3$ at 1100 °C. Using the boron:metal ratio = 6, gadolinium and cerium hexaborides MB_6 (M = Gd, Ce) were prepared without the coexisting of the corresponding tetraborides at 1200–1400 °C and 1100 °C, respectively. A small number of inclusions (oxides, borates, etc.) can be completely removed using acid treatment with conc. HCl solution [170].

Crystalline ytterbium hexaboride YbB_6 along with some amount of amorphous boron were prepared by heating ytterbium(II) *closo*-decaborate $Yb[B_{10}H_{10}]$ in a quartz tube maintained at 10^{-5} Torr to a maximum of 1000 °C [171].

The thermal decomposition of copper(I) *closo*-decaborane $Cu_2[B_{10}H_{10}]$ at 800 °C was found to produce crystalline copper boride CuB_{24} and metal copper and amorphous boron [172].

Recently, the annealing of copper(II) complexes with hydrazine $[Cu^{II}(N_2H_4)_3][B_{10}H_{10}]\cdot nH_2O$ or ammonia $[Cu^{II}(NH_3)_4][B_{10}H_{10}]\cdot nH_2O$ in argon at 900 °C was used to prepare a Cu@BN boron-containing copper composite [173]. The composition consists of a boron nitride matrix doped with cubic copper(0) nanoparticles with an average particle size of ~81 nm or ~52 nm, respectively.

Modern technology has a high demand for materials which can operate under extremal temperatures. Inorganic polymers attract attention because they offer some properties that are not found in organic materials, such as low-temperature flexibility, electrical conductivity, and nonflammability. The linear polysilicates obtained by the polycondensation of sodium metasilicate with silanol groups are the most widely studied among the non-organic polymers [174].

Prior to studying the distribution of salts of the boron cluster anions in the silicate matrix, the thermal and thermomechanical properties of starting salts $(R_3NH)_2[B_{12}H_{12}]$ (R = Et, Bu) were examined as compared with $(Et_3NH)_2[B_{10}H_{10}]$ [175]. The TGA and DSC data for $(R_3NH)_2[B_{12}H_{12}]$ are similar; thermal destruction is observed at 260–450 °C, and the weakening of intermolecular contacts (softening) is observed before thermooxidative destruction. As for $(Et_3NH)_2[B_{10}H_{10}]$, thermooxidative and thermal destructions occur simultaneously within a narrow temperature range of 260–320 °C, and the softening temperature lies within the range of intensive weight loss.

Furthermore, we studied the thermal behavior of triethylammonium *closo*-decaborate in a silicate matrix [176]. The interaction of sodium silicates of liquid glass (LG) with triethylammonium salts of boron cluster anions was studied in a wide range of component ratios. The compositions formed by addition of different amounts of $(Et_3NH)_2[B_{10}H_{10}]$ (5, 15, 30, 40, 50, 60, and 74 wt%) into sodium liquid glass [176] were studied.

The dissolution of triethylammonium salts of boron cluster anions in sodium LG at room temperature is accompanied by the release of triethylamine, which completely stops when the temperature rises to 100 °C. The absence of a band of stretching vibrations of the NH groups of the triethylammonium cation in the region of 3100–3200 cm^{-1} indicates its complete replacement in the composition by Na^+ ions. The retention of the formed sodium salts in the silicate matrix is carried out due to the formation of specific cation–anion contacts.

It was found that for compositions with *closo*-decaborate anion, the anion oxidation in air begins at 350 °C and is accompanied by a significant exothermic effect. IR spectroscopic analysis of the thermolysis products obtained in air at 350 and 600 °C showed the presence of the *closo*-decaborate anion in the samples [176,177]. A branched 3D system of multicenter bonds between BH-groups of the boron cluster and silanol groups via the water molecules can be assumed in the resulting inorganic polymer composition (Figure 10). The participation of the boron cluster in hydride–proton (dihydrogen) bonds is detected using IR spectroscopy because of the splitting of the band of stretching vibrations of the BH groups ν(BH) observed near 2500 cm^{-1}, whereas the formation of hydrogen bonds between sylanol groups and water molecules can be assumed because of the presence of broadened band ν(OH) in the region 3600–3000 cm^{-1}. This structure prevents the *closo*-

decaborate anion from undergoing complete degradation, thus forming a surface protective layer which consists of borates and silicates, allowing preventing the bulky sample from oxygen diffusion and its further oxidation at high temperatures. The authors concluded that samples are stable up to 600 °C, which is attractive for fabricating boron-rich thermally stable coatings.

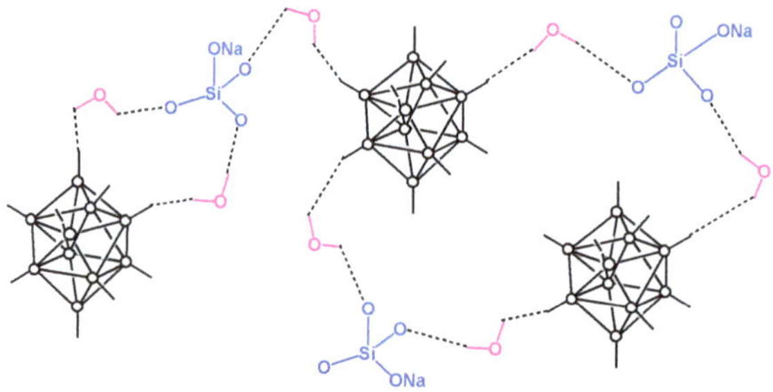

Figure 10. Multicenter interactions between $(Et_3NH)_2[B_{10}H_{10}]$ and components of sodium liquid glass.

In the IR spectra of the compositions, the multiplet splitting of the band of stretching vibrations of BH bonds, which is characteristic of interactions of this kind, is clearly manifested. The thermal stability of individual salts of boron cluster anions is determined by the nature of the anion and cation of the starting compound. The thermal stability of the compositions also depends on the nature of the boron cluster anion. According to the TG and DSC data, the protective layer is formed when the temperature rises to 500 °C during the thermogravimetric analysis. It is worth noting that is that the heat treatment of the sample under these conditions is accompanied by a high exothermic effect, which can lead to the melting of the borosilicate components.

The possibility of using compositions based on the *closo*-decaborate anion as highly heat-resistant boron-enriched materials is evidenced by TMA data [178,179]. The samples are characterized by high heat resistance compared to the original components; they do not soften at temperatures $\geq$ 600 °C as high thermal and thermomechanical stability is probably ensured due to the formation of a "protective structure" on the surface of the samples during testing, which prevents the diffusion of atmospheric oxygen.

Structural features of boron cluster anions introduced into the compositions have a significant effect on the processes and thermomechanical properties of the compositions.

The formation of a protective layer in the compositions is also observed for the *closo*-dodecaborate anion [180,181]. It was found that heating the composition, in which the amount of the doped component is 60 wt%, leads to the formation of the composite and crystallization of sodium salt of the *closo*-dodecaborate anion on its surface at 200 °C, according to X-ray powder diffraction data [182]. The results of the study of the morphology of the obtained sample by scanning electron microscopy were compared with the results of the morphology of the sodium salt obtained from an aqueous solution. On the surface of the composite, there are needle-shaped nanosized particles with well-formed faces and sizes of 60–100 nm in width and up to 3 μm in length. In a sample of sodium salt obtained from an aqueous solution, only large blocks with a size of about 10–30 μm are present. In addition, when the initial mixture contained 60% of triethylammonium *closo*-dodecaborate at 450 °C, a high plasticization of the composition was noted, as evidenced by the TMA data [181]. The obtained properties may be important for the processing of composites and are probably due to the presence of about 6.6% triethylammonium substituted derivative of the *closo*-dodecaborate anion in the reaction, which is formed during the heat treatment of the composition.

The reaction of LG with the triethylammonium salt of the perchlorinated substituted derivative of the *closo*-decaborate anion proceeds similarly [179,180]. For LG/$[B_{10}Cl_{10}]^{2-}$ compositions containing the perchlorinated *closo*-decaborate anion up to 20 wt%, their plasticizing properties were determined. In the presence of small amounts of additives, associations formed between the silicate and polyhedral boron anions, which act as crosslinking agents. According to the TMA data, triethylammonium salt of the perchlorinated *closo*-decaborate anion does not soften up to a degradation temperature of 420 °C, whereas in salt $(Et_3NH)_2[B_{10}H_{10}]$, this process occurs at 245 °C. Differences in the deformation stability are also retained in compositions containing equimolar amounts of boron cluster anions. Analyzing the obtained TMA results [179,180], it is obvious that the deformation stability of the system containing the perchlorinated anion is significantly higher compared to that of the decahydro-*closo*-decaborate anion. This fact indicates a more rigid structuring observed in the presence of the perchlorinated anion.

Compositions with a low content of boron cluster anions are of particular interest for studying the structural features of the associates formed. We suggested that the associates formed in the silicate matrix can be distributed as individual particles. which was determined using transmission electron microscopy (TEM). For samples containing the $[B_{10}H_{10}]^{2-}$ anion, the TEM image shows isolated elongated particles 12.5–47.5 nm in size. In turn, the shape of the particles for the composition with the perchlorinated anion is not so pronounced [180]. In the latter case, particles 5–40 nm in size form agglomerates distributed in a silicate matrix. Thus, it is obvious that the shape and nature of the distribution of associates formed in the silicate matrix directly depends on the nature of the boron cluster anion. As a result of these studies, we have patented a boron-containing neutron shielding material [183], which was obtained by the reaction between sodium silicate $Na_2O(SiO_2)_n$ in an aqueous solution of sodium hydroxide with trimethylammonium decahydro-*closo*-decaborate $(Me_3NH)_2[B_{10}H_{10}]$; the reaction solution was boiled until the trimethylamine formed as a result of the reaction of the sodium hydroxide solution with $(Me_3NH)_2[B_{10}H_{10}]$ is completely removed, then dried by raising the temperature to 300 °C. Due to the numerous supramolecular contacts that appear in the glass structure, the destruction of the $[B_{10}H_{10}]^{2-}$ anion is not observed up to 600 °C. In addition, the boron content in the product is from 15 to 40 wt%, which provides a high ability of the material to capture thermal neutrons. The obtained neutron-shielding material can be used, in particular, in the encapsulation of radioactive waste, in the creation of protective shields.

Boron-containing compounds can be used as light components for the creation of metal matrix composites [184,185]. The metal composites containing copper and aluminum as matrices and salts $Cs_2[B_{10}H_{10}]$, $[Me_2NH_2]_2[B_{10}H_{10}]$, $[Ph_4P]_2[B_{10}H_{10}]$, $[Et_3NH]_2[B_{10}Cl_{10}]$, $Cs_2[B_{12}H_{12}]$, $[Et_3NH]_2[B_{12}H_{12}]$ [149], and $[Bu_4N]_2[B_{12}H_{12}]$ [150] were prepared and coated onto a steel surface. It was shown that the developed metal matrix composites with the boron cluster anion salts can be applied for coatings. A friction cladding method allows one to prepare high-quality coatings, providing a high adhesion of the coating to the metal substrate. No defects were found either in the mass of the coating or on the surface.

5. Conclusions

Here, we tried to summarize briefly the synthetic routes and wide application fields of boron-containing materials prepared from boron cluster anions and carboranes. The recent renaissance in chemistry of borohydrides and carboranes is associated with ever-new prospects for their practical use. We hope that the information collected in this article will significantly expand the understanding of the variability of the practical application of boron cluster anions and carboranes to obtain composites and materials based on them, and will provide a novel perspective on the ways to obtain composites with desired properties.

Author Contributions: Conceptualization, I.B.S. and E.A.M.; resources, S.E.N.; writing—original draft preparation, V.V.A.; writing—review and editing, I.B.S.; supervision, project administration, N.T.K. All authors have read and agreed to the published version of the manuscript.

Funding: The work was performed within the framework of the State Assignment of the Kurnakov Institute of General and Inorganic Chemistry, Russian Academy of Sciences and the State Assignment of the Nesmeyanov Institute of Organoelement Compounds, Russian Academy of Sciences in the field of fundamental scientific research.

Data Availability Statement: No new data were created.

Conflicts of Interest: The authors declare no conflict of interest.

References

1. Goodger, E. Unconventional fuels. *New Sci.* **1957**, 27–29.
2. Martin, D.R. The development of borane fuels. *J. Chem. Ed.* **1959**, *36*, 208–214. [CrossRef]
3. United States Government. *Boron High Energy Fuels: Hearings Before the Committee on Science and Astronautics, U.S. House of Representatives. Eighty-Six Congress, First Session*; United States Government Printing Office: Washington, DC, USA, 1959.
4. Sivaev, I.B. Nitrogen heterocyclic salts of polyhedral borane anions: From ionic liquids to energetic materials. *Chem. Heterocycl. Comp.* **2017**, *53*, 638–658. [CrossRef]
5. Shan, Z.; Sheng, L.; Yang, R. Research progress of icosahedral polyhydroborate $B_{12}H_{12}{}^{2-}$ anion compounds and their application in propellants and explosives. *Chin. J. Explos. Propellants* **2017**, *40*, 1–16. [CrossRef]
6. Jiao, N.; Zhang, Y.; Liu, L.; Shreeve, J.M.; Zhang, S. Hypergolic fuels based on water-stable borohydride cluster anions with ultralow ignition delay times. *J. Mater. Chem. A* **2017**, *5*, 13341–13346. [CrossRef]
7. Rachiero, G.P.; Titi, H.M.; Rogers, R.D. Versatility and remarkable hypergolicity of *exo*-6, *exo*-9 imidazole-substituted *nido*-decaborane. *Chem. Commun.* **2017**, *53*, 7736–7739. [CrossRef]
8. Sheng, L.; Shan, Z.; Guo, X.; Yang, R. Synthesis, characterization and thermal behavior of bis(dialkyl-5-aminotetrazolium) dodecadodecaborates. *Chin. J. Org. Chem.* **2018**, *38*, 2093–2100. [CrossRef]
9. Zhang, Z.; Zhang, Y.; Li, Z.; Jiao, N.; Liu, L.; Zhang, S. $B_{12}H_{12}{}^{2-}$-Based metal (Cu^{2+}, Ni^{2+}, Zn^{2+}) complexes as hypergolic fuels with superior hypergolicity. *Eur. J. Inorg. Chem.* **2018**, *2018*, 981–986. [CrossRef]
10. Li, H.; Zhang, Y.; Liu, L.; Jiao, N.; Meng, X.; Zhang, S. Amino functionalized $[B_{12}H_{12}]^{2-}$ salts as hypergolic fuels. *New J. Chem.* **2018**, *42*, 3568–3573. [CrossRef]
11. Derdziuk, J.; Malinowski, P.J.; Jaroń, T. Synthesis, structural characterization and thermal decomposition studies of $(N_2H_5)_2B_{12}H_{12}$ and its solvates. *Int. J. Hydrogen Energy* **2019**, *44*, 27030–27038. [CrossRef]
12. Sharon, P.; Afri, M.; Mitlin, S.; Gottlieb, L.; Schmerling, B.; Grinstein, D.; Welner, S.; Frimer, A.A. Preparation and characterization of bis(guanidinium) and bis(aminotetrazolium)dodecahydroborate salts: Green high energy nitrogen and boron rich compounds. *Polyhedron* **2019**, *157*, 71–89. [CrossRef]
13. Li, S.; Pan, X.; Jiang, Y.; Chang, S.; Jin, X.; Yang, Y.; Huang, X.; Guo, Y. Ignition and combustion behaviors of high energetic polyhedral boron cluster. *Propellants Explos. Pyrotech.* **2019**, *44*, 1319–1326. [CrossRef]
14. Wang, Q.-Y.; Wang, J.; Wang, S.; Wang, Z.-Y.; Cao, M.; He, C.-L.; Yang, J.-Q.; Zang, S.-Q.; Mak, T.C.W. *o*-Carborane-based and atomically precise metal clusters as hypergolic materials. *J. Am. Chem. Soc.* **2020**, *142*, 12010–12014. [CrossRef]
15. Hawthorne, M.F. The role of chemistry in the development of boron neutron capture therapy of cancer. *Angew. Chem. Int. Ed.* **1993**, *32*, 950–984. [CrossRef]
16. Soloway, A.H.; Tjarks, W.; Barnum, B.A.; Rong, F.-G.; Barth, R.F.; Codogni, I.M.; Wilson, J.G. The chemistry of neutron capture therapy. *Chem. Rev.* **1998**, *98*, 1515–1562. [CrossRef]
17. Sivaev, I.B.; Bregadze, V.I. Polyhedral boranes for medical applications: Current status and perspectives. *Eur. J. Inorg. Chem.* **2009**, *2009*, 1433–1450. [CrossRef]
18. Hu, K.; Yang, Z.; Zhang, L.; Xie, L.; Wang, L.; Xu, H.; Josephson, L.; Liang, S.H.; Zhang, M.-R. Boron agents for neutron capture therapy. *Coord. Chem. Rev.* **2020**, *405*, 213139. [CrossRef]
19. Scholz, M.; Hey-Hawkins, E. Carbaboranes as pharmacophores: Properties, synthesis, and application strategies. *Chem. Rev.* **2011**, *111*, 7035–7062. [CrossRef]
20. Fink, K.; Uchman, M. Boron cluster compounds as new chemical leads for antimicrobial therapy. *Coord. Chem. Rev.* **2021**, *431*, 213684. [CrossRef]
21. Kugler, M.; Nekvinda, J.; Holub, J.; El Anwar, S.; Das, V.; Šícha, V.; Pospíšilová, K.; Fábry, M.; Král, V.; Brynda, J.; et al. Inhibitors of CA IX enzyme based on polyhedral boron compounds. *ChemBioChem* **2021**, *22*, 2741–2761. [CrossRef] [PubMed]
22. Valliant, J.F.; Guenther, K.J.; King, A.S.; Morel, P.; Schaffer, P.; Sogbein, O.O.; Stephenson, K.A. The medicinal chemistry of carboranes. *Coord. Chem. Rev.* **2002**, *232*, 173–230. [CrossRef]
23. Issa, F.; Kassiou, M.; Rendina, L.M. Boron in drug discovery: Carboranes as unique pharmacophores in biologically active compounds. *Chem. Rev.* **2011**, *111*, 5701–5722. [CrossRef]
24. Zargham, E.O.; Mason, C.A.; Lee, M.W. The use of carboranes in cancer drug development. *Int. J. Cancer Clin. Res.* **2019**, *6*, 110. [CrossRef]
25. Chen, Y.; Du, F.; Tang, L.; Xu, J.; Zhao, Y.; Wu, X.; Li, M.; Shen, J.; Wen, Q.; Cho, C.H.; et al. Carboranes as unique pharmacophores in antitumor medicinal chemistry. *Mol. Ther. Oncolytics* **2022**, *24*, 400–416. [CrossRef]

26. Avdeeva, V.V.; Garaev, T.M.; Malinina, E.A.; Zhizhin, K.Y.; Kuznetsov, N.T. Physiologically active compounds based on membranotropic cage carriers—Derivatives of adamantane and polyhedral boron clusters (Review). *Russ. J. Inorg. Chem.* **2022**, *67*, 28–47. [CrossRef]
27. Marforio, T.D.; Mattioli, E.J.; Zerbetto, F.; Calvaresi, M. Exploiting blood transport proteins as carborane supramolecular vehicles for boron neutron capture therapy. *Nanomaterials* **2023**, *13*, 1770. [CrossRef]
28. Morin, J.F.; Sasaki, T.; Shirai, Y.; Guerrero, J.M.; Tour, J.M. Synthetic routes toward carborane-wheeled nanocars. *J. Org. Chem.* **2007**, *72*, 9481–9490. [CrossRef] [PubMed]
29. Sasaki, T.; Guerrero, J.M.; Leonard, A.D.; Tour, J.M. Nanotrains and self-assembled two-dimensional arrays built from carboranes linked by hydrogen bonding of dipyridones. *Nano Res.* **2008**, *1*, 412–419. [CrossRef]
30. Chiang, P.-T.; Mielke, J.; Godoy, J.; Guerrero, J.M.; Alemany, L.B.; Villagómez, C.J.; Saywell, A.; Grill, G.; Tour, J.M. Toward a light-driven motorized nanocar: Synthesis and initial imaging of single molecules. *ACS Nano* **2012**, *6*, 592–597. [CrossRef] [PubMed]
31. Hosseini Lavasani, S.M.; Pishkenari, H.N.; Meghdari, A. How chassis structure and substrate crystalline direction affect the mobility of thermally driven *p*-carborane-wheeled nanocars. *J. Phys. Chem. C* **2019**, *123*, 4805–4824. [CrossRef]
32. Hawthorne, M.F.; Zink, J.I.; Skelton, J.M.; Bayer, M.B.; Liu, C.; Livshits, E.; Baer, R.; Neuhauser, D. Electrical or photocontrol of the rotary motion of a metallacarborane. *Science* **2004**, *303*, 1849–1852. [CrossRef] [PubMed]
33. Hawthorne, M.F.; Ramachandran, B.M.; Kennedy, R.D.; Knobler, C.B. Approaches to rotary molecular motors. *Pure Appl. Chem.* **2006**, *78*, 1299–1304. [CrossRef]
34. Shlyakhtina, N.I.; Safronov, A.V.; Sevryugina, Y.V.; Jalisatgi, S.S.; Hawthorne, M.F. Synthesis, characterization, and preliminary fluorescence study of a mixed ligand bis(dicarbollyl)nickel complex bearing a tryptophan-BODIPY FRET couple. *J. Organomet. Chem.* **2015**, *798*, 234–244. [CrossRef]
35. Sivaev, I.B. Ferrocene and transition metal bis(dicarbollides) as platform for design of rotatory molecular switches. *Molecules* **2017**, *22*, 2201. [CrossRef]
36. Anufriev, S.A.; Timofeev, S.V.; Anisimov, A.A.; Suponitsky, K.Y.; Sivaev, I.B. Bis(dicarbollide) complexes of transition metals as a platform for molecular switches. Study of complexation of 8,8′-bis(methylsulfanyl) derivatives of cobalt and iron bis(dicarbollides). *Molecules* **2020**, *25*, 5745. [CrossRef]
37. Bae, Y.S.; Farha, O.K.; Spokoyny, A.M.; Mirkin, C.A.; Hupp, J.T.; Snurr, R.Q. Carborane-based metal–organic frameworks as highly selective sorbents for CO_2 over methane. *Chem. Commun.* **2008**, *35*, 4135–4137. [CrossRef]
38. Spokoyny, A.M.; Farha, O.K.; Mulfort, K.L.; Hupp, J.T.; Mirkin, C.A. Porosity tuning of carborane-based metal–organic frameworks (MOFs) via coordination chemistry and ligand design. *Inorg. Chim. Acta* **2010**, *364*, 266–271. [CrossRef]
39. Kennedy, R.D.; Krungleviciute, V.; Clingerman, D.J.; Mondloch, J.E.; Peng, Y.; Wilmer, C.E.; Sarjeant, A.A.; Snurr, R.Q.; Hupp, J.T.; Yildirim, T.; et al. Carborane-based metal-organic framework with high methane and hydrogen storage capacities. *Chem. Mater.* **2013**, *25*, 3539–3543. [CrossRef]
40. Kennedy, R.D.; Clingerman, D.J.; Morris, W.; Wilmer, C.E.; Sarjeant, A.A.; Stern, C.L.; O'Keeffe, M.; Snurr, R.Q.; Hupp, J.T.; Farha, O.K.; et al. Metallacarborane-based metal–organic framework with a complex topology. *Cryst. Growth Des.* **2014**, *14*, 1324–1330. [CrossRef]
41. Clingerman, D.J.; Morris, W.; Mondloch, J.E.; Kennedy, R.D.; Sarjeant, A.A.; Stern, C.; Hupp, J.T.; Farha, O.K.; Mirkin, C.A. Stabilization of a highly porous metal–organic framework utilizing a carborane-based linker. *Chem. Commun.* **2015**, *51*, 6521–6523. [CrossRef]
42. Boldog, I.; Bereciartua, P.J.; Bulánek, R.; Kučeráková, M.; Tomandlová, M.; Dušek, M.; Macháček, J.; De Vos, D.; Baše, T. 10-Vertex *closo*-carborane: A unique ligand platform for porous coordination polymers. *CrystEngComm* **2016**, *18*, 2036–2040. [CrossRef]
43. Boldog, I.; Dušek, M.; Jelínek, T.; Švec, P.; de Ramos, O.F.S.; Růžička, A.; Bulánek, R. Porous 10- and 12-vertex (bi)-*p*-dicarba-*closo*-boranedicarboxylates of cobalt and their gas adsorptive properties. *Microporous Mesoporous Mater.* **2018**, *271*, 284–294. [CrossRef]
44. Zhang, Y.; Hu, J.; Krishna, R.; Wang, L.; Yang, L.; Cui, X.; Duttwyler, S.; Xing, H. Rational design of microporous MOFs with anionic boron cluster functionality and cooperative dihydrogen binding sites for highly selective capture of acetylene. *Angew. Chem. Int. Ed.* **2020**, *59*, 17664–17669. [CrossRef]
45. Sun, W.; Hu, J.; Duttwyler, S.; Wang, L.; Krishna, R.; Zhang, Y. Highly selective gas separation by two isostructural boron cluster pillared MOFs. *Sep. Purif. Technol.* **2022**, *283*, 120220. [CrossRef]
46. Li, Z.; Li, X.-B.; Light, M.E.; Carrillo, A.E.; Arauzo, A.; Valvidares, M.; Roscini, C.; Teixidor, F.; Viñas, C.; Gándara, F.; et al. A metal-organic framework incorporating eight different size rare-earth metal elements: Toward multifunctionality *À La Carte*. *Adv. Funct. Mater.* **2023**, *33*, 2307369. [CrossRef]
47. Sivaev, I.B. Decaborane: From Alfred Stock and rocket fuel projects to nowadays. *Molecules* **2023**, *28*, 6287. [CrossRef]
48. Grimes, R.N. *Carboranes*, 3rd ed.; Academic Press: London, UK, 2016. [CrossRef]
49. Sivaev, I.B.; Prikaznov, A.V.; Naoufal, D. Fifty years of the *closo*-decaborate anion chemistry. *Collect. Czechoslov. Chem. Commun.* **2010**, *75*, 1149–1199. [CrossRef]
50. Zhizhin, K.Y.; Zhdanov, A.P.; Kuznetsov, N.T. Derivatives of *closo*-decaborate anion $[B_{10}H_{10}]^{2-}$ with *exo*-polyhedral substituents. *Russ. J. Inorg. Chem.* **2010**, *55*, 2089–2127. [CrossRef]

51. Mahfouz, N.; Abi Ghaida, F.; El Hajj, Z.; Diab, M.; Floquet, S.; Mehdi, A.; Naoufal, D. Recent achievements on functionalization within *closo*-decahydrodecaborate $[B_{10}H_{10}]^{2-}$ clusters. *ChemistrySelect* **2022**, *7*, e202200770. [CrossRef]
52. Hawthorne, M.F.; Shelly, K.; Li, F. The versatile chemistry of the $[B_{20}H_{18}]^{2-}$ ions: Novel reactions and structural motifs. *Chem. Commun.* **2002**, *6*, 547–554. [CrossRef]
53. Avdeeva, V.V.; Malinina, E.A.; Zhizhin, K.Y.; Bernhardt, E.; Kuznetsov, N.T. Structural diversity of dimer clusters based on the octadecahydro-eicosaborate anion. *J. Struct. Chem.* **2019**, *60*, 692–712. [CrossRef]
54. Thompson, A.G. MOCVD technology for semiconductors. *Mater. Lett.* **1997**, *30*, 255–263. [CrossRef]
55. Jones, A.C. Recent developments in the MOCVD of electronic materials. *Adv. Mater. Opt. Electron.* **2000**, *10*, 91–92. [CrossRef]
56. Chow, L.A. Equipment and manufacturability issues in chemical vapor deposition processes. In *Handbook of Thin Film Deposition*, 4th ed.; Seshan, K., Schepis, D., Eds.; William Andrew: Oxford, UK, 2018; pp. 269–316. [CrossRef]
57. Global Metal Organic Chemical Vapor Deposition Market: Analysis by Application, by Category, by Region Size and Trends with Impact of COVID-19 and Forecast up to 2026. Available online: https://www.researchandmarkets.com/report/mocvd (accessed on 1 June 2023).
58. Roush, G. The Toxicology of the Boranes. *J. Occup. Med.* **1959**, *1*, 46–52. Available online: https://www.jstor.org/stable/44999044 (accessed on 1 June 2023). [CrossRef]
59. Naeger, L.L.; Leibman, K.C. Mechanisms of decaborane toxicity. *Toxicol. Appl. Pharmacol.* **1972**, *22*, 517–527. [CrossRef]
60. Nakamura, K. Preparation and properties of amorphous boron films deposited by pyrolysis of decaborane in the molecular flow region. *J. Electrochem. Soc.* **1984**, *131*, 2691. [CrossRef]
61. Kamimura, K.; Yoshimura, T.; Nagaoka, T.; Nakao, M.; Onuma, Y.; Makimura, M. Preparation and thermoelectric property of boron thin film. *J. Solid State Chem.* **2000**, *154*, 153–156. [CrossRef]
62. Natsir, M.; Sagara, A.; Motojima, O. Reduction of hydrogen content in boron film by controlling glow discharge conditions. *J. Nucl. Mater.* **1995**, *220–222*, 865–868. [CrossRef]
63. Kodama, H.; Oyaidzu, M.; Sasaki, M.; Kimura, H.; Morimoto, Y.; Oya, Y.; Matsuyama, M.; Sagara, A.; Noda, N.; Okuno, K. Studies on structural and chemical characterization for boron coating films deposited by PCVD. *J. Nucl. Mater.* **2004**, *329–333*, 889–893. [CrossRef]
64. Saidoh, M.; Ogiwara, N.; Shimada, M.; Arai, T.; Hiratsuka, H.; Koike, T.; Shimizu, M.; Ninomiya, H.; Nakamura, H.; Jimbou, R. Initial boronization of JT-60U tokamak using decaborane. *Jpn. J. Appl. Phys.* **1993**, *32*, 3276–3281. [CrossRef]
65. Perkins, F.K.; Rosenberg, R.A.; Lee, S.; Dowben, P.A. Synchrotron-radiation-induced deposition of boron and boron carbide films from boranes and carboranes: Decaborane. *J. Appl. Phys.* **1991**, *69*, 4103–4109. [CrossRef]
66. Kim, Y.G.; Dowben, P.A.; Spencer, J.T.; Ramseyer, G.O. Chemical vapor deposition of boron and boron nitride from decaborane(14). *J. Vac. Sci. Technol. A* **1989**, *7*, 2796–2799. [CrossRef]
67. Chatterjee, S.; Luo, Z.; Acerce, M.; Yates, D.M.; Johnson, A.T.C.; Sneddon, L.G. Chemical vapor deposition of boron nitride nanosheets on metallic substrates via decaborane/ammonia reactions. *Chem. Mater.* **2011**, *23*, 4414–4416. [CrossRef]
68. Chatterjee, S.; Kim, M.J.; Zakharov, D.N.; Kim, S.M.; Stach, E.A.; Maruyama, B.; Sneddon, L.G. Syntheses of boron nitride nanotubes from borazine and decaborane molecular precursors by catalytic chemical vapor deposition with a floating nickel catalyst. *Chem. Mater.* **2012**, *24*, 2872–2879. [CrossRef]
69. Bellott, B.J.; Noh, W.; Nuzzo, R.G.; Girolami, G.S. Nanoenergetic materials: Boron nanoparticles from the pyrolysis of decaborane and their functionalisation. *Chem. Commun.* **2009**, *22*, 3214–3215. [CrossRef] [PubMed]
70. Ekimov, E.A.; Zibrov, I.P.; Zoteev, A.V. Preparation of boron microcrystals via high-pressure, high-temperature pyrolysis of decaborane, $B_{10}H_{14}$. *Inorg. Mater.* **2011**, *47*, 1194–1198. [CrossRef]
71. Ekimov, E.A.; Lebed, J.B.; Sidorov, V.A.; Lyapin, S.G. High-pressure synthesis of crystalline boron in B-H system. *Sci. Technol. Adv. Mater.* **2011**, *12*, 055009. [CrossRef]
72. Kher, S.S.; Spencer, J.T. Chemical vapor deposition precursor chemistry. 3. Formation and characterization of crystalline nickel boride thin films from the cluster-assisted deposition of polyhedral borane compounds. *Chem. Mater.* **1992**, *4*, 538–544. [CrossRef]
73. Tynell, T.; Aizawa, T.; Ohkubo, I.; Nakamura, K.; Mori, T. Deposition of thermoelectric strontium hexaboride thin films by a low pressure CVD method. *J. Cryst. Growth* **2016**, *449*, 10–14. [CrossRef]
74. Kher, S.S.; Tan, Y.; Spencer, J.T. Chemical vapor deposition of metal borides. 4. The application of polyhedral boron clusters to the chemical vapor deposition formation of gadolinium boride thin-film materials. *Appl. Organomet. Chem.* **1996**, *10*, 297–304. [CrossRef]
75. Kher, S.S.; Romero, J.V.; Caruso, J.D.; Spencer, J.T. Chemical vapor deposition of metal borides. 6. The formation of neodymium boride thin film materials from polyhedral boron clusters and metal halides by chemical vapor deposition. *Appl. Organomet. Chem.* **2008**, *22*, 300–307. [CrossRef]
76. Guélou, G.; Martirossyan, M.; Ogata, K.; Ohkubo, I.; Kakefuda, Y.; Kawamoto, N.; Kitagawa, Y.; Ueda, J.; Tanabe, S.; Maeda, K.; et al. Rapid deposition and thermoelectric properties of ytterbium boride thin films using hybrid physical chemical vapor deposition. *Materialia* **2018**, *1*, 244–248. [CrossRef]
77. Ohkubo, I.; Aizawa, T.; Nakamura, K.; Mori, T. Control of competing thermodynamics and kinetics in vapor phase thin-film growth of nitrides and borides. *Front. Chem.* **2021**, *9*, 642388. [CrossRef] [PubMed]
78. Su, K.; Sneddon, L.G. A polymer precursor route to metal borides. *Chem. Mater.* **1993**, *5*, 1659–1668. [CrossRef]

79. Pender, M.J.; Sneddon, L.G. An efficient template synthesis of aligned boron carbide nanofibers using a single-source molecular precursor. *Chem. Mater.* **2000**, *12*, 280–283. [CrossRef]
80. Borchardt, L.; Kockrick, E.; Wollmann, P.; Kaskel, S.; Guron, M.M.; Sneddon, L.G.; Geiger, D. Ordered mesoporous boron carbide based materials *via* precursor nanocasting. *Chem. Mater.* **2010**, *22*, 4660–4668. [CrossRef]
81. Wei, X.; Carroll, P.J.; Sneddon, L.G. Ruthenium-catalyzed ring-opening polymerization syntheses of poly(organodecaboranes): New single-source boron-carbide precursors. *Chem. Mater.* **2006**, *18*, 1113–1123. [CrossRef]
82. Seyferth, D.; Rees, W.S.; Haggerty, J.S.; Lightfoot, A. Preparation of boron-containing ceramic materials by pyrolysis of the decaborane(14)-derived [-$B_{10}H_{12}$·Ph_2POPPh_2]$_x$- polymer. *Chem. Mater.* **1989**, *1*, 45–52. [CrossRef]
83. Yu, X.-H.; Cao, K.; Huang, Y.; Yang, J.; Li, J.; Chang, G. Platinum catalyzed sequential hydroboration of decaborane: A facile approach to poly(alkenyldecaborane) with decaborane in the main chain. *Chem. Commun.* **2014**, *50*, 4585–4587. [CrossRef] [PubMed]
84. Wang, J.; Gou, Y.; Jian, K.; Huang, J.; Wang, H. Boron carbide ceramic hollow microspheres prepared from poly(6-CH_2=$CH(CH_2)_4$-$B_{10}H_{13}$) precursor. *Mater. Des.* **2016**, *109*, 408–414. [CrossRef]
85. Welna, D.T.; Bender, J.D.; Wei, X.; Sneddon, L.G.; Allcock, H.R. Preparation of boron-carbide/carbon nanofibers from a poly(norbornenyldecaborane) single-source precursor *via* electrostatic spinning. *Adv. Mater.* **2005**, *17*, 859–862. [CrossRef]
86. Li, J.; Yang, H.; Liang, Y.; Chen, J.; Xie, M.; Zhou, L.; Xiong, Q.; Li, W. Ruthenium-catalyzed cascade ring-opening polymerization/hydroboration for the synthesis of low cross-linking poly(6-norbornenyldecaborane) and its thermal property. *Adv. Mater. Res.* **2014**, *1035*, 288–291. [CrossRef]
87. Zhang, X.; Li, J.; Cao, K.; Yi, Y.; Yang, J.; Li, B. Synthesis and characterization of B-C polymer hollow microspheres from a new organodecaborane preceramic polymer. *RSC Adv.* **2015**, *5*, 86214–86218. [CrossRef]
88. Wang, J.; Gou, Y.; Zhang, Q.; Jian, K.; Chen, Z.; Wang, H. Linear organodecaborane block copolymer as a single-sourceprecursor for porous boron carbide ceramics. *J. Eur. Ceram. Soc.* **2017**, *37*, 1937–1943. [CrossRef]
89. Li, J.; Cao, K.; Li, J.; Liu, M.; Zhang, S.; Yang, J.; Zhang, Z.; Li, B. Synthesis and ceramic conversion of a new organodecaborane preceramic polymer with high-ceramic-yield. *Molecules* **2018**, *23*, 2461. [CrossRef]
90. Pender, M.J.; Wideman, T.; Carroll, P.J.; Sneddon, L.G. Transition metal promoted reactions of boron hydrides. 15. Titanium-catalyzed decaborane–olefin hydroborations: One-step, high-yield syntheses of monoalkyldecaboranes. *J. Am. Chem. Soc.* **1998**, *120*, 9108–9109. [CrossRef]
91. Mazurowski, J.; Lee, S.; Ramseyer, G.; Dowben, P.A. Characterization of boron carbide films formed by PECVD. *MRS Online Proc. Libr.* **1992**, *242*, 637–642. [CrossRef]
92. Lee, S.; Mazurowski, J.; Ramseyer, G.; Dowben, P.A. Characterization of boron carbide thin films fabricated by plasma enhanced chemical vapor deposition from boranes. *J. Appl. Phys.* **1992**, *72*, 4925–4933. [CrossRef]
93. Lee, S.; Mazurowski, J.; O'Brien, W.L.; Dong, Q.Y.; Jia, J.J.; Callcott, T.A.; Tan, Y.; Miyano, K.E.; Ederer, D.L.; Mueller, D.R.; et al. The structural homogeneity of boron carbide thin films fabricated using plasma enhanced chemical vapor deposition from B_5H_9+CH_4. *J. Appl. Phys.* **1993**, *74*, 6919–6924. [CrossRef]
94. Gold, J.W.; Militscher, C.; Slauson, D.D. Pentaborane Disposal: Taming the Dragon. Available online: https://dokumen.tips/documents/pentaborane-taming-the-dragonpdf.html (accessed on 1 June 2023).
95. Rees, W.S.; Seyferth, D. High-yield synthesis of B_4C/BN ceramic materials by pyrolysis of polymeric Lewis base adducts of decaborane(14). *J. Am. Ceram. Soc.* **1988**, *71*, C194–C196. [CrossRef]
96. Rees, W.S.; Seyferth, D. Preparation, characterization, and pyrolysis of decaborane (14)—Based polymers: B_4C/BN and BN procedures. In *A Collection of Papers Presented at the 13th Annual Conference on Composites and Advanced Ceramic Materials: Ceramic Engineering and Science Proceedings*; The American Ceramic Society: Westerville, OH, USA, 1989; Volume 10, pp. 837–845. [CrossRef]
97. Yogo, T.; Naka, S. Synthesis of boron nitride from triammoniadecaborane and hydrazine under pressure. *J. Mater. Sci.* **1990**, *25*, 374–378. [CrossRef]
98. Yogo, T.; Naka, S.; Iwahara, H. Synthesis of cubic boron nitride with boron nitride powder formed from triammoniadecaborane. *J. Mater. Sci.* **1991**, *26*, 3758–3762. [CrossRef]
99. Robertson, B.W.; Adenwalla, S.; Harken, A.; Welsch, P.; Brand, J.I.; Dowben, P.A.; Claassen, J.P. A class of boron-rich solid-state neutron detectors. *Appl. Phys. Lett.* **2002**, *80*, 3644–3646. [CrossRef]
100. Harken, A.D.; Day, E.E.; Robertson, B.W.; Adenwalla, S. Boron-rich semiconducting boron carbide neutron detector. *Jpn. J. Appl. Phys.* **2005**, *44*, 444–445. [CrossRef]
101. Day, E.; Diaz, M.J.; Adenwalla, S. Effect of bias on neutron detection in thin semiconducting boron carbide films. *J. Phys. D* **2006**, *39*, 2920–2924. [CrossRef]
102. Caruso, A.N.; Dowben, P.A.; Balkir, S.; Schemm, N.; Osberg, K.; Fairchild, R.W.; Flores, O.B.; Balaz, S.; Harken, A.D.; Robertson, B.W.; et al. The all boron carbide diode neutron detector: Comparison with theory. *Mater. Sci. Eng. B* **2006**, *135*, 129–133. [CrossRef]
103. Hong, N.; Crow, L.; Adenwalla, S. Time-of-flight neutron detection using PECVD grown boron carbide diode detector. *Nucl. Instrum. Methods Phys. Res. Sect. A* **2013**, *708*, 19–23. [CrossRef]
104. Peterson, G.; Su, Q.; Wang, Y.; Dowben, P.A.; Nastasi, M. Improved p–n heterojunction device performance induced by irradiation in amorphous boron carbide films. *Mater. Sci. Eng. B* **2015**, *202*, 25–30. [CrossRef]

105. Nastasi, M.; Peterson, G.; Su, Q.; Wang, Y.; Ianno, N.J.; Benker, N.; Echeverría, E.; Yost, J.A.; Kelber, A.J.; Dong, B.; et al. Electrical and structural characterization of neutron irradiated amorphous boron carbide/silicon p-n heterojunctions. *Nucl. Inst Methods Phys. Res. B* **2018**, *432*, 48–54. [CrossRef]
106. Byun, D.; Spady, B.R.; Ianno, N.J.; Dowben, P.A. Comparison of different chemical vapor deposition methodologies for fabrication of hetero junction boron-carbide diodes. *Nanostruct. Mater.* **1995**, *5*, 465–471. [CrossRef]
107. Zhang, D.; Mcilroy, D.N.; O'Brien, W.L.; De Stasio, G. The chemical and morphological properties of boron-carbon alloys grown by plasma-enhanced chemical vapour deposition. *J. Mater. Sci.* **1998**, *33*, 4911–4915. [CrossRef]
108. Nordell, B.J.; Karki, S.; Nguyen, T.D.; Rulis, P.; Caruso, A.N.; Purohit, S.S.; Li, H.; King, S.W.; Dutta, D.; Gidley, D.; et al. The influence of hydrogen on the chemical, mechanical, optical/electronic, and electrical transport properties of amorphous hydrogenated boron carbide. *J. Appl. Phys.* **2015**, *118*, 035703. [CrossRef]
109. Bute, A.; Jagannath; Kar, R.; Chopade, S.S.; Desai, S.S.; Deo, M.N.; Rao, P.; Chand, N.; Kumar, S.; Singh, K.; et al. Effect of self-bias on the elemental composition and neutron absorption of boron carbide films deposited by RF plasma enhanced CVD. *Mater. Chem. Phys.* **2016**, *182*, 62–71. [CrossRef]
110. Nordell, B.J.; Keck, C.L.; Nguyen, T.D.; Caruso, A.N.; Purohit, S.S.; Lanford, W.A.; Dutta, D.; Gidley, D.; Henry, P.; King, S.W.; et al. Tuning the properties of a complex disordered material: Full factorial investigation of PECVD-grown amorphous hydrogenated boron carbide. *Mater. Chem. Phys.* **2016**, *173*, 268–284. [CrossRef]
111. Bute, A.; Jena, S.; Bhattacharya, D.; Kumar, S.; Chand, N.; Keskar, N.; Sinha, S. Composition dependent microstructure and optical properties of boron carbide (B_xC) thin films deposited by radio frequency-plasma enhanced chemical vapour deposition technique. *Mater. Res. Bull.* **2019**, *109*, 175–182. [CrossRef]
112. Tu, R.; Hu, X.; Li, J.; Yang, M.; Li, Q.; Shi, J.; Li, H.; Ohmori, H.; Goto, T.; Zhang, S. Fabrication of (a-nc) boron carbide thin films via chemical vapor deposition using *ortho*-carborane. *J. Asian Ceram. Soc.* **2020**, *8*, 327–335. [CrossRef]
113. Bute, A.; Jena, S.; Kedia, S.; Udupa, D.V.; Singh, K.; Bhattacharya, D.; Modi, M.H.; Chand, N.; Sinha, S. Boron carbide thin films deposited by RF-PECVD and PLD technique: A comparative study based on structure, optical properties, and residual stress. *Mater. Chem. Phys.* **2021**, *258*, 123860. [CrossRef]
114. Bute, A.; Jena, S.; Sharma, R.K.; Jagannath, U.D.V.; Maiti, N. Linear and non-linear optical properties of boron carbide thin films. *Appl. Surf. Sci.* **2023**, *608*, 155101. [CrossRef]
115. Caruso, A.N.; Billa, R.B.; Balaz, S.; Brand, J.I.; Dowben, P.A. The heteroisomeric diode. *J. Phys. Condens. Matter.* **2004**, *16*, L139–L146. [CrossRef]
116. Caruso, A.N.; Balaz, S.; Xu, B.; Dowben, P.A.; McMullen-Gunn, A.S.; Brand, J.I.; Losovyj, Y.B.; McIlroy, D.N. Surface photovoltage effects on the isomeric semiconductors of boron-carbide. *Appl. Phys. Lett.* **2004**, *84*, 1302–1304. [CrossRef]
117. Echeverría, E.; Dong, B.; Liu, A.; Wilson, E.R.; Peterson, G.; Nastasi, M.; Dowben, P.A.; Kelber, J.A. Strong binding at the gold (Au) boron carbide interface. *Surf. Coat. Technol.* **2017**, *314*, 51–54. [CrossRef]
118. Caruso, A.N. The physics of solid-state neutron detector materials and geometries. *J. Phys. Condens. Matter* **2010**, *22*, 443201. [CrossRef]
119. Peterson, G.G.; Echeverria, E.; Dong, B.; Silva, J.P.; Wilson, E.R.; Kelber, J.A.; Nastasi, M.; Dowben, P.A. Increased drift carrier lifetime in semiconducting boron carbides deposited by plasma enhanced chemical vapor deposition from carboranes and benzene. *J. Vac. Sci. Technol. A* **2017**, *35*, 03E101. [CrossRef]
120. Oyelade, A.; Yost, A.J.; Benker, N.; Dong, B.; Knight, S.; Schubert, M.; Dowben, P.A.; Kelber, J.A. Composition-dependent charge transport in boron carbides alloyed with aromatics: Plasma enhanced chemical vapor deposition aniline/*ortho*-carborane films. *Langmuir* **2018**, *34*, 12007–12016. [CrossRef]
121. Dong, B.; James, R.; Kelber, J.A. PECVD of boron carbide/aromatic composite films: Precursor stability and resonance stabilization energy. *Surf. Coat. Technol.* **2016**, *290*, 94–99. [CrossRef]
122. Dong, B.; Oyelade, A.; Nandagopal, N.; Kelber, J.A. Aromatic-doped boron carbide films formed by PECVD of metacarborane and aniline or pyridine: Chemical and electronic structure. *Surf. Coat. Technol.* **2017**, *314*, 45–50. [CrossRef]
123. Echeverría, E.; James, R.; Chiluwal, U.; Pasquale, F.L.; Colón Santana, J.A.; Gapfizi, R.; Tae, J.-D.; Driver, M.S.; Enders, A.; Kelber, J.A.; et al. Novel semiconducting boron carbide/pyridine polymers for neutron detection at zero bias. *Appl. Phys. A* **2015**, *118*, 113–118. [CrossRef]
124. Echeverria, E.; Dong, B.; Peterson, G.; Silva, J.P.; Wilson, E.R.; Driver, M.S.; Jun, Y.-S.; Stucky, G.D.; Knight, S.; Hofmann, T.; et al. Semiconducting boron carbides with better charge extraction through the addition of pyridine moieties. *J. Phys. D* **2016**, *49*, 355302. [CrossRef]
125. Oyelade, A.; Osonkie, A.; Yost, A.J.; Benker, N.; Dowben, P.A.; Kelber, J.A. Optical, electronic and visible-range photo-electronic properties of boron carbide-indole films. *J. Phys. D* **2020**, *53*, 355101. [CrossRef]
126. Hwang, S.-D.; Remmes, N.B.; Dowben, P.A.; McIlroy, D.N. Nickel doping of boron carbide grown by plasma enhanced chemical vapor deposition. *J. Vac. Sci. Technol. B* **1996**, *14*, 2957–2960. [CrossRef]
127. Hwang, S.-D.; Remmes, N.; Dowben, P.A.; McIlroy, D.N. Nickel doping of boron-carbon alloy films and corresponding Fermi level shifts. *J. Vac. Sci. Technol. B* **1997**, *15*, 854–859. [CrossRef]
128. Hwang, S.-D.; Yang, K.; Dowben, P.A.; Ahmad, A.A.; Ianno, N.J.; Li, J.Z.; Lin, J.Y.; Jiang, H.X.; McIlroy, D.N. Fabrication of *n*-type nickel doped $B_5C_{1+\delta}$ homojunction and heterojunction diodes. *Appl. Phys. Lett.* **1997**, *70*, 1028–1030. [CrossRef]

129. Carlson, L.; LaGraffe, D.; Balaz, S.; Ignatov, A.; Losovyj, Y.B.; Choi, J.; Dowben, P.A.; Brand, J.I. Doping of boron carbides with cobalt, using cobaltocene. *Appl. Phys. A* **2007**, *89*, 195–201. [CrossRef]
130. Ignatov, A.Y.; Losovyj, Y.B.; Carlson, L.; LaGraffe, D.; Brand, J.I.; Dowben, P.A. Pairwise cobalt doping of boron carbides with cobaltocene. *J. Appl. Phys.* **2007**, *102*, 083520. [CrossRef]
131. Dowben, P.A.; Kizilkaya, O.; Liu, J.; Montag, B.; Nelson, K.; Sabirianov, I.; Brand, J.I. 3d transition metal doping of semiconducting boron carbides. *Mater. Lett.* **2009**, *63*, 72–74. [CrossRef]
132. Sharapov, V.M.; Mirnov, S.V.; Grashin, S.A.; Lebedev, S.V.; Kovan, I.A.; Krasilnikov, A.V.; Krupin, V.A.; Levin, L.S.; Romannikov, A.N.; Zakharov, A.P. Boronization of Russian tokamaks from carborane precursors. *J. Nucl. Mater.* **1995**, *220–222*, 730–735. [CrossRef]
133. Hong, S.H.; Lee, K.S.; Kim, K.M.; Kim, H.T.; Kim, G.P.; Sun, J.H.; Woo, H.J.; Park, J.M.; Kim, W.C.; Kim, H.K.; et al. KSTAR team. First boronization in KSTAR. *Fusion Eng. Des.* **2010**, *85*, 946–949. [CrossRef]
134. Hong, S.-H.; Lee, K.-S.; Kim, K.-P.; Kim, K.-M.; Kim, H.-T.; Sun, J.-H.; Woo, H.-J.; Park, J.-M.; Park, F.-K.; Kim, W.-C.; et al. First boronization in KSTAR: Experiences on carborane. *J. Nucl. Mater.* **2011**, *415*, S1050–S1053. [CrossRef]
135. Ko, J.; Den Hartog, D.J.; Goetz, J.A.; Weix, P.J.; Limbach, S.T. First gaseous boronization during pulsed discharge cleaning. *J. Nucl. Mater.* **2013**, *432*, 146–151. [CrossRef]
136. Sharapov, V.M. Discharge chamber plasma-chemical conditioning in magnetic confinement fusion devices (Review). *Phys. Atom. Nucl.* **2021**, *84*, 1266–1271. [CrossRef]
137. Zheng, H.; Thorne, K.; Mackenzie, J.D.; Yang, X.; Hawthorne, M.F. Boron carbide-based ceramics via polymer route synthesis. *MRS Online Proc. Libr.* **1991**, *249*, 15–23. [CrossRef]
138. Johnson, S.E.; Yang, X.; Hawthorne, M.F.; Thorne, K.J.; Zheng, H.; Mackenzie, J.D. Alkynyl substituted carboranes as precursors to boron carbide thin films, fibers and composites. *MRS Online Proc. Libr.* **1992**, *271*, 833–838. [CrossRef]
139. Cheng, S.; Yuan, K.; Wang, X.; Han, J.; Jian, X.; Wang, J. Poly(phenylene-carborane) for boron-carbide/carbon ceramic precursor synthesized *via* nickel catalysis. *Polymer* **2017**, *115*, 224–231. [CrossRef]
140. Yan, D.; Chen, J.; Zhang, Y.; Gou, Y. Synthesis and characterization of a carborane-containing precursor for B_4C ceramics. *Sci. Discov.* **2021**, *9*, 128–132. [CrossRef]
141. Yan, D.; Chen, J.; Zhang, Y.; Gou, Y. Preparation of novel carborane-containing boron carbide precursor and its derived ceramic hollow microsphere. *Ceram. Int.* **2022**, *48*, 18392–18400. [CrossRef]
142. Filonenko, V.P.; Zinin, P.V.; Zibrov, I.P.; Anokhin, A.S.; Kukueva, E.V.; Lyapin, S.G.; Fominski, V.Y. Synthesis of star-shaped boron carbide micro-crystallites under high pressure and high temperatures. *Crystals* **2018**, *8*, 448. [CrossRef]
143. Pavlov, I.S.; Ivanova, A.G.; Filonenko, V.P.; Zibrov, I.P.; Voloshin, A.E.; Zinin, P.V.; Vasiliev, A.L. The rhombic hexecontahedron-boron carbide microcrystals—Crystal structure analysis. *Scr. Mater.* **2023**, *222*, 115023. [CrossRef]
144. Bagramov, R.H.; Filonenko, V.P.; Zibrov, I.P.; Skryleva, E.A.; Nikolaev, A.V.; Pasternak, D.G.; Vlasov, I.I. Highly boron-doped graphite and diamond synthesized from adamantane and *ortho*-carborane under high pressure. *Materialia* **2022**, *21*, 101274. [CrossRef]
145. Avdeeva, V.V.; Malinina, E.A.; Kuznetsov, N.T. Boron cluster anions and their derivatives in complexation reactions. *Coord. Chem. Rev.* **2022**, *469*, 214636. [CrossRef]
146. Matveev, E.Y.; Avdeeva, V.V.; Zhizhin, K.Y.; Malinina, E.A.; Kuznetsov, N.T. Effect of nature of substituents on coordination properties of mono- and disubstituted derivatives of boron cluster anions $[B_nH_n]^{2-}$ (n = 10, 12) and carboranes with *exo*-polyhedral B-X Bonds (X = N, O, S, Hal). *Inorganics* **2022**, *10*, 238. [CrossRef]
147. Paskevicius, M.; Hansen, B.R.S.; Jørgensen, M.; Richter, B.; Jensenet, T.R. Multifunctionality of silver *closo*-boranes. *Nat. Commun.* **2017**, *8*, 15136. [CrossRef] [PubMed]
148. Zhao, X.; Yang, Z.; Chen, H.; Wang, Z.; Zhou, X.; Zhang, H. Progress in three-dimensional aromatic-like *closo*-dodecaborate. *Coord. Chem. Rev.* **2021**, *444*, 214042. [CrossRef]
149. Avdeeva, V.V.; Malinina, E.A.; Kuznetsov, N.T. Isomerism in salts and complexes with boron cluster anions $[B_{10}H_{10}]^{2-}$ and $[B_{20}H_{18}]^{2-}$. *Russ. J. Inorg. Chem.* **2020**, *65*, 335–358. [CrossRef]
150. Avdeeva, V.V.; Malinina, E.A.; Kuznetsov, N.T. Isomerism in complexes with the decahydro-*closo*-decaborate anion. *Polyhedron* **2016**, *105*, 205–221. [CrossRef]
151. Kravchenko, E.A.; Gippius, A.A.; Tkachev, A.V.; Zhurenko, S.V.; Golubev, A.V.; Kubasov, A.S.; Selivanov, N.A.; Buzanov, G.A.; Bykov, A.Y.; Zhizhin, K.Y.; et al. Non-covalent Cl···X interactions in several silver compounds based on $[B_{12}Cl_{12}]^{2-}$ clusters: ^{35}Cl NQR and X-ray diffraction. *Inorg. Chim. Acta* **2023**, *544*, 121231. [CrossRef]
152. Avdeeva, V.V.; Kravchenko, E.A.; Gippius, A.A.; Vologzhanina, A.V.; Malinina, E.A.; Zhurenko, S.V.; Buzanov, G.A.; Kuznetsov, N.T. Decachloro-*closo*-decaborate anion in copper(II) complexation reactions with N-donor ligands: ^{35}Cl NQR and X-ray studies. *Polyhedron* **2017**, *127*, 238–247. [CrossRef]
153. Albert, B.; Hofmann, K. Metal Borides: Versatile Structures and Properties. In *Handbook of Solid State Chemistry*; John Wiley & Sons: Hoboken, NJ, USA, 2017. [CrossRef]
154. Avdeeva, V.V.; Buzin, M.I.; Dmitrienko, A.O.; Dorovatovskii, P.V.; Malinina, E.A.; Kuznetsov, N.T.; Voronova, E.D.; Zubavichus, Y.V.; Vologzhanina, A.V. Solid-state reactions of eicosaborate $[B_{20}H_{18}]^{2-}$ salts and complexes. *Chem. Eur. J.* **2017**, *23*, 16819–16828. [CrossRef]

155. Eleazer, B.J.; Smith, M.D.; Peryshkov, D.V. Reaction of a ruthenium B-carboranyl hydride complex and $BH_3(SMe_2)$: Selective formation of a pincer-supported metallaborane $LRu(B_3H_8)$. *Tetrahedron* **2019**, *75*, 1471–1474. [CrossRef]
156. Korolenko, S.E.; Zhuravlev, K.P.; Tsaryuk, V.I.; Kubasov, A.S.; Avdeeva, V.V.; Malinina, E.A.; Burlov, A.S.; Divaeva, L.N.; Zhizhin, K.Y.; Kuznetsov, N.T. Crystal structures, luminescence, and DFT study of mixed-ligand Zn(II) and Cd(II) complexes with phenyl-containing benzimidazole derivatives with linker CN or NN group. *J. Lumin.* **2021**, *237*, 118156. [CrossRef]
157. Malinina, E.A.; Vologzhanina, A.V.; Avdeeva, V.V.; Goeva, L.V.; Efimov, N.N.; Ugolkova, E.A.; Minin, V.V.; Kuznetsov, N.T. Structures, magnetic properties, and EPR studies of tetranuclear copper(II) complexes $[Cu_4(OH)_4L_4]^{4+}$ (L = bpa, bipy) stabilized by anions containing decahydro-*closo*-decaborate anion. *Polyhedron* **2020**, *183*, 114540. [CrossRef]
158. Chung, H.-Y.; Weinberger, M.B.; Levine, J.B.; Kavner, A.; Yang, J.-M.; Tolbert, S.H.; Kaner, R.B. Synthesis of ultra-incompressible superhard rhenium diboride at ambient pressure. *Science* **2007**, *316*, 436–439. [CrossRef]
159. Latini, A.; Rau, J.V.; Teghil, R.; Generosi, A.; Albertini, V.R. Superhard properties of rhodium and iridium boride films. *ACS Appl. Mater. Interfaces* **2010**, *2*, 581–587. [CrossRef]
160. Xie, M.; Mohammadi, R.; Turner, C.L.; Kaner, R.B.; Kavner, A.; Tolbert, S.H. Lattice stress states of superhard tungsten tetraboride from radial X-ray diffraction under nonhydrostatic compression. *Phys. Rev. B Condens. Matter Mater. Phys.* **2014**, *90*, 104104. [CrossRef]
161. Bazhin, P.; Chizhikov, A.; Bazhina, A.; Konstantinov, A.; Avdeeva, V. Titanium-titanium boride matrix composites prepared in-situ under conditions combining combustion processes and high-temperature shear deformation. *Mater. Sci. Eng. A* **2023**, *874*, 145093. [CrossRef]
162. Prokopets, A.D.; Bazhin, P.M.; Konstantinov, A.S.; Antipov, M.S.; Avdeeva, V.V. Structural features of layered composite material $TiB_2/TiAl/Ti_6Al_4$ obtained by unrestricted SHS-compression. *Mater. Lett.* **2021**, *300*, 130165. [CrossRef]
163. Petrichko, M.I.; Karavaev, I.A.; Savinkina, E.V.; Grigoriev, M.S.; Buzanov, G.A.; Retivov, V.M. Rare-earth nitrate complexes with dimethylformamide. *Russ. J. Inorg. Chem.* **2023**, *68*, 415–423. [CrossRef]
164. Avdeeva, V.V.; Polyakova, I.N.; Vologzhanina, A.V.; Goeva, L.V.; Buzanov, G.A.; Generalova, N.B.; Malinina, E.A.; Zhizhin, K.Y.; Kuznetsov, N.T. $[Co(solv)_6][B_{10}H_{10}]$ (solv = DMF and DMSO) for low-temperature synthesis of borides. *Russ. J. Inorg. Chem.* **2016**, *61*, 1125–1134. [CrossRef]
165. Malinina, E.A.; Myshletsov, I.I.; Buzanov, G.A.; Kubasov, A.S.; Kozerozhets, I.V.; Goeva, L.V.; Nikiforova, S.E.; Avdeeva, V.V.; Zhizhin, K.Y.; Kuznetsov, N.T. A New approach to the synthesis of nanocrystalline cobalt boride in the course of the thermal decomposition of cobalt complexes $[Co(DMF)_6]^{2+}$ with boron cluster anions. *Molecules* **2023**, *28*, 453. [CrossRef] [PubMed]
166. Avdeeva, V.V.; Vologzhanina, A.V.; Ugolkova, E.A.; Minin, V.V.; Malinina, E.A.; Kuznetsov, N.T. Synthesis and structures of compounds $[ML_6][B_{10}Cl_{10}]$ (M = Co, Ni; L = CH_3CN, DMF, DMSO) as precursors for synthesis of cobalt(II) and nickel(II) complexes with organic L ligands. *J. Solid State Chem.* **2021**, *296*, 121989. [CrossRef]
167. Malinina, E.A.; Goeva, L.V.; Buzanov, G.A.; Avdeeva, V.V.; Efimov, N.N.; Kozerozhets, I.V.; Kuznetsov, N.T. Synthesis and physicochemical properties of binary cobalt(II) borides. Thermal reduction of precursor complexes $[CoL_n][B_{10}H_{10}]$ (L = H_2O, $n = 6$; N_2H_4, $n = 3$). *Russ. J. Inorg. Chem.* **2019**, *64*, 1325–1334. [CrossRef]
168. Malinina, E.A.; Goeva, L.V.; Buzanov, G.A.; Avdeeva, V.V.; Efimov, N.N.; Kuznetsov, N.T. A new method for synthesis of binary borides with desired properties. *Dokl. Chem.* **2019**, *487*, 180–183. [CrossRef]
169. Malinina, E.A.; Goeva, L.V.; Buzanov, G.A.; Retivov, V.M.; Avdeeva, V.V.; Kuznetsov, N.T. Synthesis and thermal reduction of complexes $[NiL_n][B_{10}H_{10}]$ (L = DMF, H_2O, $n = 6$; L = N_2H_4, $n = 3$): Formation of solid solutions $Ni_3C_{1-x}B_x$. *Russ. J. Inorg. Chem.* **2020**, *65*, 126–132. [CrossRef]
170. Itoh, H.; Tsuzuki, Y.; Yogo, T.; Naka, S. Synthesis of cerium and gadolinium borides using boron cage compounds as a boron source. *Mater. Res. Bull.* **1987**, *22*, 1259–1266. [CrossRef]
171. White III, J.P.; Shore, S.G. Complexes of divalent lanthanides (ytterbium(II), europium(II), and samarium(II)) with decaborates. *Inorg. Chem.* **1992**, *31*, 2756–2761. [CrossRef]
172. Volkov, V.V.; Yur'ev, G.S.; Solomatina, L.Y.; Voronina, G.S.; Myakishev, K.G. Thermal changes in decahydro-*closo*-decaborate(2-)copper(I) $Cu_2B_{10}H_{10}$ and the nature of copper borides. *Bull. Acad. Sci. USSR Div. Chem. Sci.* **1990**, *39*, 435–438. [CrossRef]
173. Malinina, E.A.; Myshletsov, I.I.; Buzanov, G.A.; Kozerozhets, I.V.; Simonenko, N.P.; Simonenko, T.L.; Nikiforova, S.E.; Avdeeva, V.V.; Zhizhin, K.Y.; Kuznetsov, N.T. Physicochemical fundamentals of the synthesis of a Cu@BN composite consisting of nanosized copper enclosed in a boron nitride matrix. *Inorganics* **2023**, *11*, 345. [CrossRef]
174. Skachkova, V.K.; Grachev, A.V.; Shaulov, A.Y.; Berlin, A.A. Inorganic polymers using sodium silicate liquid glass. Features of silicate polycondensation. *Russ. J. Phys. Chem. B* **2019**, *13*, 849–852. [CrossRef]
175. Skachkova, V.K.; Goeva, L.V.; Grachev, A.V.; Avdeeva, V.V.; Malinina, E.A.; Shaulov, A.Y.; Berlin, A.A.; Kuznetsov, N.T. Thermal and thermomechanical properties of trialkylammonium dodecahydro-*closo*-dodecaborates $(R_3NH)_2[B_{12}H_{12}]$ (R = Et, Bu). *Russ. J. Inorg. Chem.* **2017**, *62*, 84–89. [CrossRef]
176. Goeva, L.V.; Skachkova, V.K.; Avdeeva, V.V.; Malinina, E.A.; Grachev, A.V.; Shaulov, A.Y.; Berlin, A.A.; Kuznetsov, N.T. Interactions of sodium liquid glass with triethylammonium decahydro-*closo*-decaborate $(Et_3NH)_2B_{10}H_{10}$. *Russ. J. Inorg. Chem.* **2014**, *59*, 107–110. [CrossRef]
177. Skachkova, V.K.; Goeva, L.V.; Grachev, A.V.; Avdeeva, V.V.; Malinina, E.A.; Shaulov, A.Y.; Berlin, A.A.; Kuznetsov, N.T. Thermal oxidation of the decahydro-*closo*-decaborate anion $B_{10}H_{10}{}^{2-}$ in a silicate matrix. *Inorg. Mater.* **2015**, *51*, 498–502. [CrossRef]

178. Skachkova, V.K.; Malinina, E.A.; Goeva, L.V.; Grachev, A.V.; Avdeeva, V.V.; Kozerozhets, I.V.; Shaulov, A.Y.; Berlin, A.A.; Kuznetsov, N.T. Morphology of supramolecular structures based on sodium silicates and $[B_{10}R_{10}]^{2-}$ (R = H, Cl) boron cluster anions. *Inorg. Mater.* **2021**, *57*, 1173–1177. [CrossRef]
179. Malinina, E.; Skachkova, V.; Goeva, L.; Grachev, A.; Lyubimova, G.; Avdeeva, V.; Kozerozhets, I.; Shaulov, A.; Berlin, A.; Kuznetsov, N. Thermomechanical properties of compositions based on polysilicates modified with boron cluster anions or SiO_2 nanoparticles. *Bol. Soc. Esp. Ceram. Vidr.* **2020**, *59*, 201–208. [CrossRef]
180. Skachkova, V.K.; Malinina, E.A.; Goeva, L.V.; Grachev, A.V.; Avdeeva, V.V.; Shaulov, A.Y.; Berlin, A.A.; Kuznetsov, N.T. Polycondensation of water glass sodium silicates in the presence of $[B_nX_n]^{2-}$ (n = 10, 12; X = H, Cl) boron cluster anions. *Inorg. Mater.* **2020**, *56*, 657–661. [CrossRef]
181. Skachkova, V.K.; Goeva, L.V.; Grachev, A.V.; Kochneva, I.K.; Malinina, E.A.; Shaulov, A.Y.; Berlin, A.A.; Kuznetsov, N.T. Composites based on triethylammonium dodecahydro-*closo*-dodecaborate ($(Et_3NH)_2[B_{12}H_{12}]$) and sodium silicate water glass. *Inorg. Mater.* **2017**, *53*, 207–211. [CrossRef]
182. Malinina, E.A.; Skachkova, V.K.; Kozerozhets, I.V.; Avdeeva, V.V.; Goeva, L.V.; Buzanov, G.A.; Shaulov, A.Y.; Berlin, A.A.; Kuznetsov, N.T. Formation of nanoscale sodium dodecahydro-*closo*-dodecaborate $Na_2[B_{12}H_{12}]$ on the surface of a silicate matrix. *Dokl. Chem.* **2019**, *484*, 1–4. [CrossRef]
183. Goeva, L.V.; Malinina, E.A.; Avdeeva, V.V.; Kuznetsov, N.T.; Skachkova, V.K.; Shaulov, A.Y.; Grachev, A.V.; Berlin, A.A. Boron-Containing Neutron Shielding Material. RF Patent RU2550156C1, 10 May 2016.
184. Popov, V.A.; Zhizhin, K.Y.; Malinina, E.A.; Ketsko, V.A.; Kuznetsov, N.T. A possibility of using mechanical alloying for developing metal matrix composites with light-weight reinforcements. *J. Alloys Compd.* **2007**, *434–435*, 451–454. [CrossRef]
185. Popov, V.A.; Zhizhin, K.Y.; Kuznetsov, N.T.; Staudhammer, K.P.; Retivov, V.M. Investigation of the possibility of application of boron clusters in composite materials with metal matrix. *Adv. Mater. Res.* **2009**, *59*, 96–100. [CrossRef]

Disclaimer/Publisher's Note: The statements, opinions and data contained in all publications are solely those of the individual author(s) and contributor(s) and not of MDPI and/or the editor(s). MDPI and/or the editor(s) disclaim responsibility for any injury to people or property resulting from any ideas, methods, instructions or products referred to in the content.

materials

MDPI

Article

Transition Metal Borides for All-in-One Radiation Shielding

Celal Avcıoğlu [1,*] and Suna Avcıoğlu [2]

1 Fachgebiet Keramische Werkstoffe/Chair of Advanced Ceramic Materials, Institute of Material Science and Technology, Faculty III Process Sciences, Technische Universität Berlin, Straße des 17, Juni 135, 10623 Berlin, Germany
2 Department of Metallurgical and Materials Engineering, Faculty of Chemistry and Metallurgy, Yıldız Technical University, 34956 Istanbul, Turkey
* Correspondence: celal.avcioglu@ceramics.tu-berlin.de

Abstract: All-in-one radiation shielding is an emerging concept in developing new-generation radiation protection materials since various forms of ionizing radiation, such as neutrons and gamma rays, can occur simultaneously. In this study, we examine the ability of transition metal borides to attenuate both photon and particle radiation. Specifically, fourteen different transition metal borides (including inner transition metal borides) are selected for examination based on their thermodynamic stabilities, molecular weights, and neutron capture cross-sections of the elements they contain. Radiation shielding characteristics of the transition metal borides are computationally investigated using Phy-X/PSD, EpiXS and NGCal software. The gamma-ray shielding capabilities of the transition metal borides are evaluated in terms of the mass attenuation coefficient (μ_m), the linear attenuation coefficient (μ), the effective atomic number (Z_{eff}), the half-value layer (HVL), the tenth-value layer (TVL), and the mean free path (MFP). The mass and linear attenuation factors are identified for thermal and fast neutrons at energies of 0.025 eV and 4 MeV, respectively. Moreover, the fast neutron removal cross-sections ($\sum R$) of the transition metal borides are calculated to assess their neutron shielding abilities. The results revealed that borides of transition metals with a high atomic number, such as Re, W, and Ta, possess outstanding gamma shielding performance. At 4 MeV photon energy, the half-value layers of ReB_2 and WB_2 compounds were found as 1.38 cm and 1.43 cm, respectively. Most notably, these HVL values are lower than the HVL value of toxic Pb (1.45 cm at 4 MeV), which is one of the conventional radiation shielding materials. On the other hand, SmB_6 and DyB_6 demonstrated exceptional neutron attenuation for thermal and fast neutrons due to the high neutron capture cross-sections of Sm, Dy, and B. The outcomes of this study reveal that transition metal borides can be suitable candidates for shielding against mixed neutron and gamma radiation.

Keywords: radiation shielding; mixed gamma-neutron radiation; transition metal borides; attenuation coefficient; simulation; Phy-X/PSD; NGCal; EpiXS

1. Introduction

Ionizing radiation is a form of energy that has a wide range of practical uses in different fields, such as power generation, healthcare, agriculture, and research [1–4]. Nevertheless, exposure to ionizing radiation can have significantly negative impacts on human health because ionizing radiation is powerful enough to detach electrons from atoms [5]. Therefore, the use of effective shielding materials is crucial to minimize human exposure to ionizing radiation.

Boron-containing materials are widely used in radiation shielding applications due to boron's high neutron capture cross-section. Boron has two stable isotopes, ^{10}B and ^{11}B. Natural occurrence rates of ^{10}B and ^{11}B isotopes are 19.1–20.3% and 79.7–80.9%, respectively [6,7]. Thermal neutron capture cross-section (3838 barns; 1 barn per nucleus = 1×10^{-24} cm^2) of the ^{10}B isotope is significantly higher than that of boron's other isotopes [8]. Therefore, ^{10}B enriched boron carbide pellets are used in control rods of nuclear reactors [9]. Borax-containing water is used in fuel storage ponds of boiling water-based nuclear reactors to draw heat away from the reactor core. Further, the aqueous solution containing sodium borax also prevents reactor corrosion. Boron-doped steel and concrete are used in the construction of nuclear power plants [8]. In addition to the nuclear energy industry, boron also finds its applications in other radiation-related fields, such as nuclear medicine, where boron is used in neutron capture therapy applications (BNCT) [10,11].

Recently, polymer matrix composites as radiation shields received a lot of attention due to their lightweight and flexibility [7,12–19]. Boron carbide is one of the favorable reinforcement materials for radiation-shielding polymer matrix composites because of its high ^{10}B content. The influence of matrix type, filler ratio, and distribution in the matrix phase and reinforcement's particle size on the composite's shielding performance has been widely studied [6,12,20–22]. Boron carbide-reinforced polymer matrix composites show excellent neutron shielding ability. Nevertheless, they exhibit poor gamma and X-ray shielding performance due to the low molecular weight of boron carbide [7,23,24].

Elements with high atomic numbers tend to provide effective shielding against high-energy radiation such as X-rays and gamma-rays. Indeed, lead, with a high density (11.5 g/cm^3) and atomic number (Z = 82), is a traditional material used for gamma-ray protection, especially in nuclear power plants, medical diagnosis and treatment centers. However, due to its high toxicity, lead has to be covered with structural materials such as concrete, which increases the thickness of the shielding material [25]. Therefore, the use of lead in nuclear shielding applications is reducing, and the plates made of W, Al, Fe, and Cu metals are emerging alternatives. In concrete structures reinforced with metal plates, metal plates provide gamma protection, while concrete provides neutron protection with its high hydrogen and oxygen content. There are also cases where concrete is enriched with boron-containing elements for advanced neutron protection [26]. However, heavy materials such as lead, concrete, or metal plates are not suitable materials for every application. Especially in agriculture, gamma and X-rays are used for pest control, improving soil and water quality, and promoting plant growth. Therefore, unlike heavy concrete and metal plates, lighter materials with even optical transmittance are required [27]. Materials that are expected to have a radiation-absorbing effect as well as optical transmittance, are also needed in spare parts of nuclear medicine and various characterization devices [28]. Intensive efforts have been made in developing new generation glass and glass-ceramic materials, especially for such applications [29]. The effectiveness of various oxide compounds, including ZnO, CuO, Dy_2O_3, Al_2O_3, V_2O_5, SiO_2, CdO, SrO, Bi_2O_3, CoO, and Nd_2O_3, as gamma-shielding additives in glass compositions has been intensively investigated due to their high molecular weight and ability to participate in the glass network structure [28,30–37].

Most materials used in the nuclear shielding industry today only provide good protection against either particle or photon type of radiation. Multi-layer radiation shields are commonly used to shield against both particle and photon radiation. In these multi-layer shields, one-layer shields against neutrons, and the other layer attenuates the gamma rays [38–40]. Nevertheless, stacking multiple layers results in thick, high-weight, and costly structures.

To address these issues, new shielding materials with good neutron and gamma-ray attenuation capacities should be explored. Transition metal borides are a class of materials with fascinating properties such as superconductivity, high hardness, high melting temperatures, and ultra incompressibility [41]. Transition metal borides show high stoichiometric and structural flexibility (M_2B, MB, MB_4, MB_6, M_3B_2, MB_2, M_2B_4, etc.). Transition metal borides can be classified according to their boron-to-metal elemental ratio. Typically, boron-rich transition metal borides display a B:M ratio of at least 4:1 whereas compounds with lower boron-to-metal ratio are known as metal-rich borides. Borides containing group 4 and 5 transition metals are ultra-refractory materials due to their melting temperatures higher than 3000 °C and their ability to maintain their hardness up to 2000 °C [42]. Borides also exhibit high strength, chemical stability, thermal conductivity, electrical conductivity, wear and corrosion resistance [43]. The discovery of the superconductivity of magnesium diboride (MgB2) led researchers to investigate borides with a similar crystal structure as MgB2 and increased the interest in metal borides. It has been reported that the planar structure of hexagonally stacked boron atoms has a great influence on the superconductivity in MB2 [44]. Recently, borides of Nb, Mo, W, and Re were also found to display superconductivity.

Transition metal borides are the combination of high-atomic-weight elements and boron, which have high neutron capture cross-section. Furthermore, there are elements with a very large neutron absorption cross-sectional area among the f-block elements, which are a subset of transition metals and are frequently referred to as lanthanides, rare-earth elements, or inner transition metals. Therefore, it is anticipated that transition metal borides may possess good attenuation capacity for both particle and photon radiations. In this contribution, for the first time in the literature, the radiation shielding performance of transition metal boride compounds is comparatively investigated.

2. Materials and Methods

2.1. Compound Selection

Two main parameters were considered to design an all-in-one radiation shielding compound. The compound should contain at least one high atomic number non-toxic element to achieve effective gamma and X-ray radiation shielding. The second criterion is the presence of at least one element with a high neutron absorption cross-section in the compound. Transition metal borides can be formed in a wide range of stoichiometric ratios and fulfill both requirements. To evaluate the shielding performance of transition metal borides, a wide range of elements with moderate to high atomic numbers (Z) were selected, and selected elements were emphasized in the periodic table in Figure 1a. Only thermodynamically stable compounds were selected. Borides of group 3 inner transition metals can form in hexa-boride (TMB_6) with cubic crystal structure and Pm3¯m symmetry [45]. In a hexa-boride crystal structure, TM is bonded in a 24-coordinate geometry to twenty-four equivalent B atoms. TM–B bond lengths are in the range of 3.05 (for LaB_6) -2.99 (for LuB_6) Å. B is bonded in a 5-coordinate geometry to four equivalent TM and five equivalent B atoms. There is one short and four long B–B bonds. The length of short B-B bonds varies in the range of 1.66 (for LaB_6) to 1.62 Å (for LuB_6). Long B-B bonds vary from 1.76 (for LaB_6) to 1.74 (for LuB_6) [45,46]. On the other hand, di-boride (TMB_2) stoichiometry with hexagonal crystal structure and P6/mmm symmetry is thermodynamically more stable for groups 4, 5, and 6 transition metal borides [47]. TM is bonded to twelve equivalent B^{3-} atoms to form a mixture of edge and face-sharing TMB_{12} cuboctahedra. All B–B bond lengths are the same, and W–B bond lengths vary in a wide range. The schematic drawings of crystal structures of di-borides and hexa-borides are presented in Figure 1b,c. Further details of transition metal borides, such as electronic state, phonon dispersion, diffraction patterns, aqueous stability, charge density, and other properties, can be found in the Materials Explorer application, which is shared openly in the public domain [48]. Therefore, hexa-borides of group 3 transition metals and di-borides of groups 4, 5, and 6 transition metal borides were selected to investigate and presented in Table 1.

Table 1. Theoretical density values of the investigated compounds [48].

				Di-Borides				
	TiB_2	ZrB_2	NbB_2	MoB_2	HfB_2	TaB_2	WB_2	ReB_2
Density (g/cm^3)	4.52	6.07	6.78	7.33	11.24	12.27	12.42	12.62
				Hexa-Borides				
	LaB_6	NdB_6	SmB_6	EuB_6	DyB_6	ErB_6		
Density (g/cm^3)	4.72	4.93	5.14	5.01	5.54	5.70		

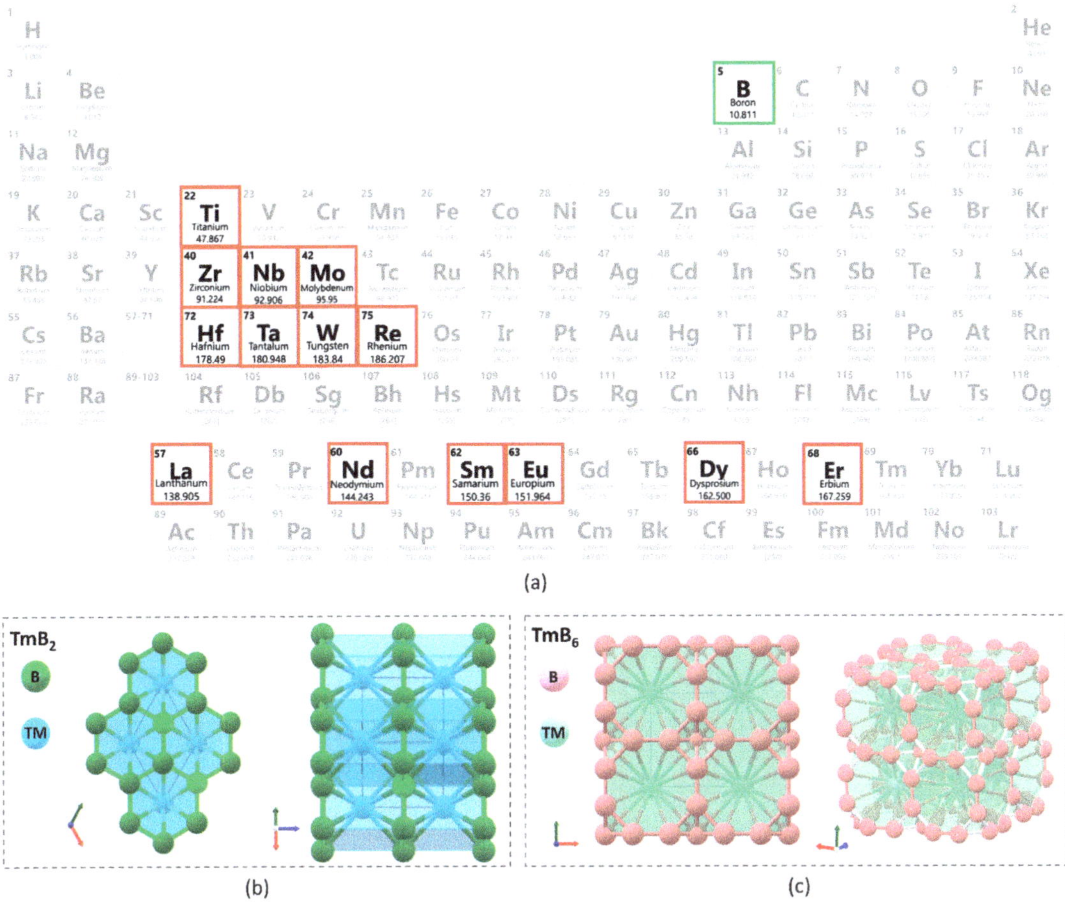

Figure 1. (**a**) Selected elements, crystal structures of (**b**) di-boride and (**c**) hexa-boride [45].

2.2. Theoretical Calculations

Theoretical simulations were conducted using PHY-X/PSD, NGCal and EpiXS software, which are freely available in the public domains for academic use [49–53]. The density values of the investigated compounds listed in Table 1 were used for calculations. The calculations were conducted based on the molar ratios of boron and transition metals in each compound. The ratio of boron to transition metal in di-boride compounds is 2:1, and boron to inner transition metal in hexa-borides is 6:1.

Firstly, the linear attenuation coefficients (LAC) and mass attenuation coefficients (MAC) were investigated. For a particular target, the LAC expresses the interaction chance between gamma rays and the target's per unit thickness. It can be defined by Lambert–Beer law, as shown in Equation (1).

$$I = I_0 e^{-\mu x} \quad (1)$$

I_0 and I represent the gamma beam's initial and attenuated intensity values. x is the target's thickness in cm, and μ is the LAC (cm^{-1}). The mass attenuation coefficient (μ_m; cm^2/g) can be obtained by dividing the LAC by the target's density (ρ), equivalent to Equation (2).

$$\mu_m = (\mu/\rho) = ln\,(I_0/I)/\rho x \quad (2)$$

The half-value layer (HVL; cm) and the tenth-value layer (TVL; cm) are commonly employed parameters for describing the shielding performance of a target material. HVL and TVL specify the needed thicknesses to obtain a 50% and 90% reduction in the intensity of the initial radiation beam. Both of these values can be obtained from the linear attenuation coefficient (μ) using Equations (3) and (4).

$$\mathrm{HVL} = ln\ (2)/\mu \quad (3)$$

$$\mathrm{TVL} = ln\ (10)/\mu \quad (4)$$

Since the neutrons do not have any charge, they can pass through the electron shell of an atom and interact with its nucleus by multiple mechanisms such as scattering, nuclear fission, and neutron capture. This phenomenon complicates the determination of neutron removal cross-section in contrast to gamma photons. Nevertheless, calculating a fast neutron removal cross-section by using Equation (5) is an effective approach to gaining knowledge about the neutron shielding capability of matter. $\sum_R$ is the total neutron removal cross-section, ρ_i is the partial density, and $(\sum_R/\rho)_i$ (cm^2/g) is the ith constituent's mass removal cross-section.

$$\sum\nolimits_R = \sum \rho_i \left(\sum\nolimits_R/\rho\right)_i \quad (5)$$

3. Results and Discussion

The mass attenuation coefficient describes the material's radiation attenuation ability, and a higher coefficient implies a greater ability for gamma-ray shielding. Figure 2a,b demonstrates the relation between the mass attenuation coefficient of the composites and incident photon energy. The trend for the mass attenuation coefficient is similar for all transition metal borides, and the lowest tested energy yields individually the greatest mass attenuation coefficient values. Further, the attenuation coefficients of all investigated transition metal borides decrease rapidly with the increase in photon energy up to 0.1 MeV. Photoelectric absorption is the most prominent interaction mechanism for low-energy photons, and it depends highly on the atomic number of elements. Indeed, the likelihood of the photoelectric effect rises with increasing atomic number of the elements in the attenuator [54]. Therefore, in the low photon energy region up to 0.1 MeV, the mass attenuation coefficient values of the transition metal borides shows direct correlation with the atomic number of the transition metals and follows the order of ReB_2 > WB_2 > TaB_2 > HfB_2 > ErB_2 > DyB_6> EuB_6 > SmB_6 > NdB_6> LaB_6 > MoB_2 > NbB_2 > ZrB_2 > TiB_2. This suggests that the likelihood of atoms of ReB_2 interacting with photons is more than that of other transition metal borides. Further, for all compounds except TiB_2, an abrupt rise in the mass attenuation coefficient values at approximately 0.04–0.08 MeV was observed (Figure 2b). A photon is absorbed during the photoelectric absorption phenomenon, and an electron is ejected from the target atom. The energy of the absorbed photon is used to remove the electron from its shell, and the electron carries away all the remaining energy. However, an electron from a particular shell cannot be ejected if the photon energy falls below the binding energy of that shell. Since the K-shell electrons, which are the most tightly bound, require high energy to get ejected, characteristic "absorption edges" appear on attenuation coefficient variation versus photon energy plots [55]. So, the observed sudden rises in attenuation coefficient plots are because of the K-absorption edges of the transition metals. Since the titanium's K-absorption edge is at 0.004966 MeV and the plots given in Figure 2a–d start from 0.01 MeV, it could not be seen in these plots. Nevertheless, the K-absorption edge of TiB_2 is visible in Figure 2e [56].

In the intermediate energy level between 0.1 MeV and 3 MeV, the reduction in mass attenuation coefficients is slowed down for all compounds. In this region, Compton scattering is the dominant interaction mechanism, and it does not heavily depend on the atomic number of the elements. Therefore, the difference between the mass attenuation coefficient values of the investigated transition metal borides is markedly reduced. In the following energy region (beyond 3 MeV), the pair production phenomenon becomes dominant over Compton scattering. Unlike Compton scattering, the pair production cross-section is proportional to the squared atomic number Z^2 of the scattering nucleus [57]. Due to this reason, the mass attenuation coefficient of the transition metal borides slowly increases above the photon energy range of 3 MeV.

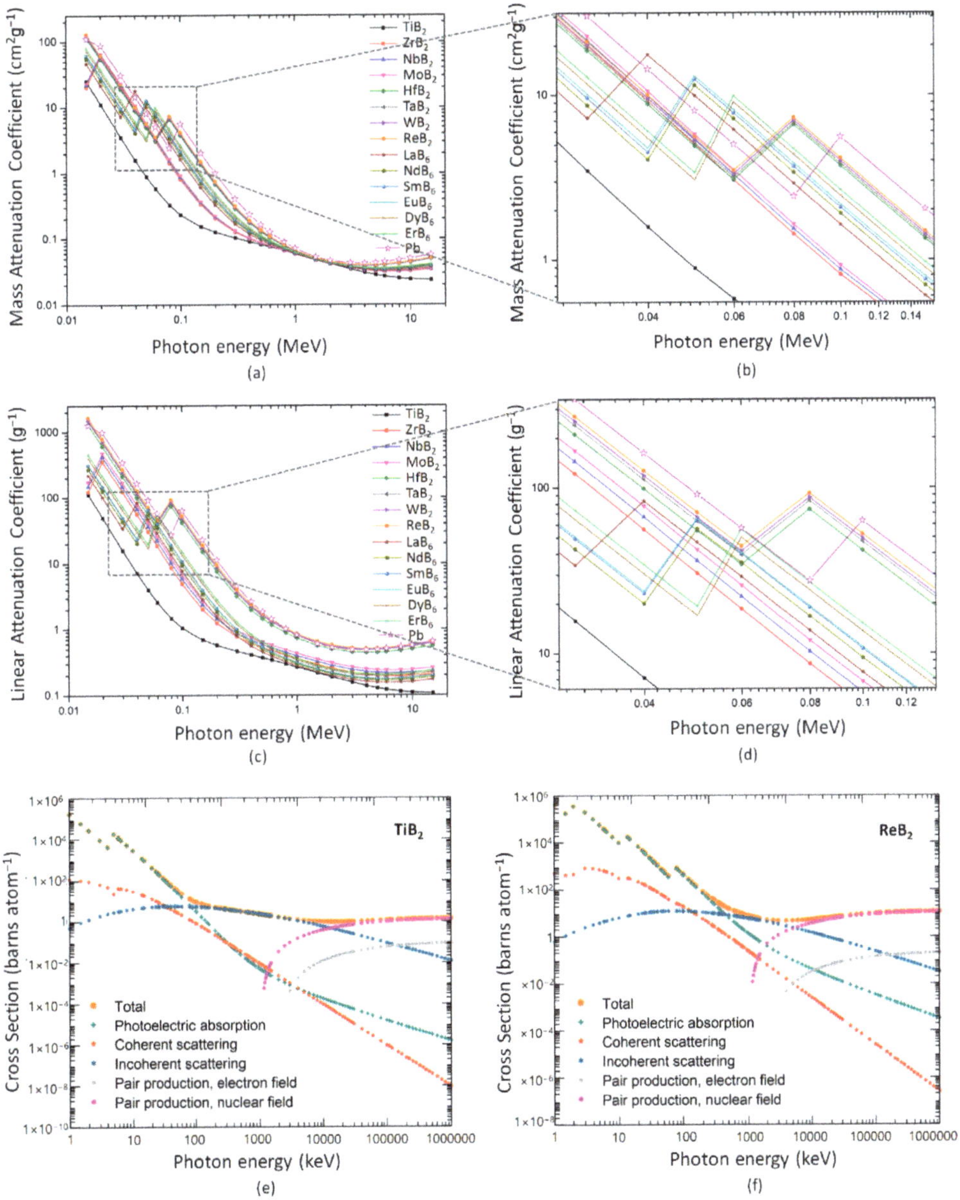

Figure 2. The variation in (**a**,**b**) the mass attenuation coefficient and (**c**,**d**) the linear attenuation coefficient with the incident photon energy for the transition metal borides and Pb. The total cross-section variations in (**e**) TiB2 and (**f**) ReB2 with the incident photon energy.

We remark that the trends for linear and mass attenuation coefficients are expectedly similar, as seen in Figure 2c. Nevertheless, unlike the mass attenuation coefficient, the density influences the linear attenuation coefficient [58]. Therefore, the difference between the linear attenuation coefficient values of the transition metal borides is higher than their mass attenuation coefficient values (Figure 2b,d). Furthermore, the linear and mass attenuation coefficient variations in ReB_2, WB_2, and TaB_2 are competitive with Pb in the selected energy region (Figure 2a–d).

The attenuation coefficient is the most crucial factor determining how gamma radiation penetrates and diffuses through extended media. The quantity of the attenuation coefficient depends on the photon energy E and the atomic number Z of the medium. Since the attenuation coefficient is proportional to the total photon interaction cross-section, the sum of the cross-sections for all the elementary scattering and absorption processes must be considered. The total cross-section can be considered as a sum of the photoelectric absorption, Compton collision, and pair production cross-sections per atom of a compound. The total cross-section variations in TiB_2 and ReB_2, which are the lowest and highest-performing compounds against photons, are presented separately in Figures 2e and 2f, respectively. The dominant regions for photoelectric absorption, Compton collision, and pair production are seen in these figures. Most notably, the incoherent scattering dominates the larger photon energy range for the TiB_2 compound in contrast to the ReB_2. Because of Re's higher Z value, ReB_2 exhibits larger photoelectric absorption and pair production cross-sections and provides better shielding than TiB_2.

Z_{eff} of the investigated transition metal borides against incident photon energies are illustrated in Figure 3. Generally, a material with a higher Z_{eff} provides more targets to collide with protons than a material with a lower Z_{eff}; hence, it experiences more interactions with photons. Therefore, a material with a high Z_{eff} value is preferred for nuclear shielding applications. It is now well established that the compounds containing heavy atoms possess high Z_{eff} values. Indeed, ReB_2, WB_2, TaB_2, and HfB_2 containing heavy transition metals exhibit comparatively higher Z_{eff} values than other samples against all photon energies.

On the other hand, the trend observed in Z_{eff} is similar to the mass attenuation coefficient, Z_{eff} of all samples first decreases then increases with the increase in photon energy. This is because of the dominating photoelectric effect, Compton scattering, and pair production in different energy zones. The Z_{eff} values of compounds lie between 12 and 74. The highest Z_{eff} values for all samples were obtained at 0.02 MeV due to the photoelectric absorption, and those are 74, 73, 72, and 71 for ReB_2, WB_2, TaB_2, and HfB_2, respectively. Contrarily, the lowest Z_{eff} values occurred in the intermediate energy region where Compton scattering is dominant (0.3–4 MeV). Although the pair production event requires a minimum energy of 1.022 MeV, the dominance of this event usually begins after photon energies of 4 MeV. Therefore, a slight increase in the Z_{eff} values beyond 4 MeV can be ascribed to the pair formation event.

The effective atomic number variation of Pb against the photon energy is also seen in Figure 3. In contrast to the investigated boride compounds, the effective atomic number of Pb is almost constant against the whole selected energy region. The linear Z_{eff} value variation indicates the stability of the target materials' shielding performance with increased photon energy [50]. The observed high Z_{eff} value change in boride compounds is because of the differences between the atomic number of boron and transition metal elements.

The mean free path, the half-value layer, and the tenth-value layer are important shielding parameters that can be used to determine the material's amount and cost for providing sufficient radiation shielding. The mean free path is the average distance a photon travels in the radiation shield before it experiences a change in its energy or direction. On the other hand, the value layer and the tenth-value layer refer to the material thickness to lower the intensity of incident radiation by half and 90%, respectively. Therefore, the lower mean free path, half-value layer, and tenth-value layer values of a shielding material correspond to greater attenuating ability. The mean free path of the investigated transition metal borides for energies up to 15 MeV is given in Figure 4a. Obviously, the mean free path of the transition metal borides is low for low-energy photons. Nevertheless, the mean free path of transition metal borides demonstrates an increasing trend with the energy increment up to approximately 4 MeV. Then, it starts to decrease above the aforementioned energy due to the pair production effect. Indeed, the highest mean free path values for all transition metal borides are observed at approximately 4 MeV. Notably, the estimated mean free path of ReB_2, WB_2, TaB_2, and HfB_2 at all tested energies are significantly lower than other transition metal borides, which can be ascribed to the higher density of these compounds as compared to the other investigated transition metal borides. Meanwhile, TiB_2 possesses the highest mean free path values at all tested photon energies due to the low atomic number of Ti.

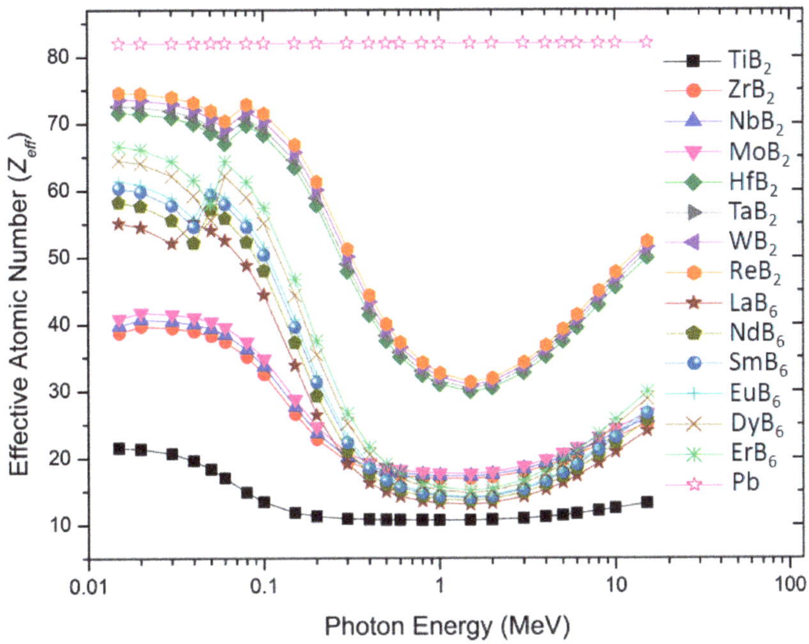

Figure 3. Variation in effective atomic numbers of the transition metal borides in the photon energy range of 0.015–15 MeV.

On the other hand, a similar trend to that of the mean free path was observed in the variation in both the half-value layer and the tenth-value layer with the incident energy. Indeed, ReB_2, WB_2, TaB_2, and HfB_2 demonstrated significantly less half-value layer and tenth-value layer values, as seen in Figure 4c,e. In these plots, ReB_2 has the lowest half-value layer and tenth-value layer values. At 4 MeV, the half-value layer and the tenth-value layer of ReB_2 reach their maximum, with 1.38 cm and 4.59 cm, respectively. Note that despite its high toxicity, lead is the most commonly used material in nuclear shielding due to its excellent gamma shielding ability. Therefore, the mean free path, half-value layer, and tenth-value layer values of ReB_2, WB_2, TaB_2, and HfB_2 are compared with those of lead, and the results are presented in Figure 3b,d,f. It can be seen that ReB_2, WB_2, and TaB_2 have better radiation shielding capacities relative to lead in the selected energy regions. It is notable to mention that Re is one of the rarest metals with an approximate abundance of 1 μg kg^{-1} [59]. On the contrary, the earth's crust contains 10^3 times higher W and Ta than Re. The results indicate that WB_2 and TaB_2 performed considerably similar shielding performance with ReB_2. So, to fabricate all-in-one radiation shielding materials, choosing WB_2 and TaB_2 over ReB_2 would be more cost-effective and sustainable.

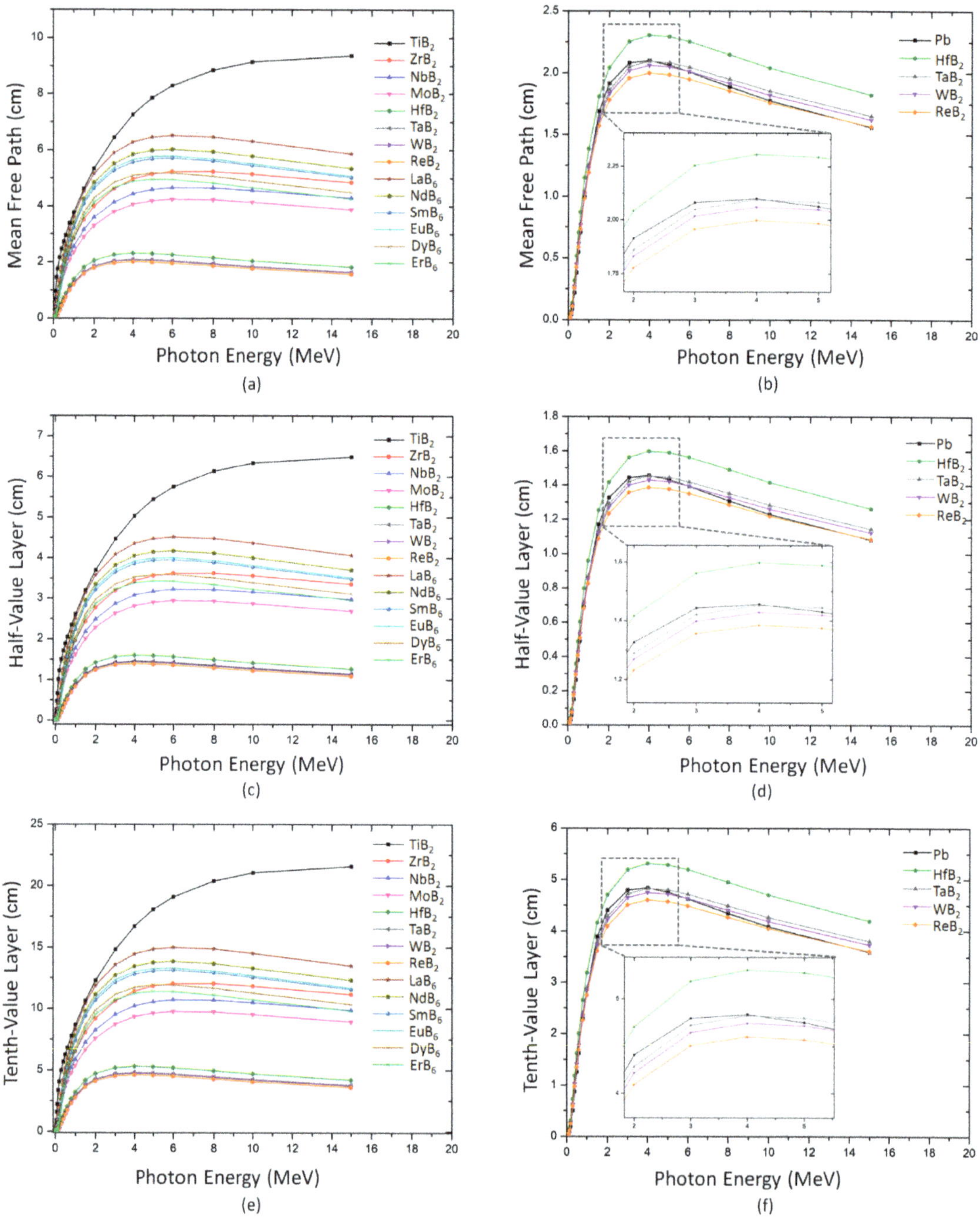

Figure 4. The variation in (**a**) the mean free path, (**c**) the half-value layer, and (**e**) the tenth-value layer with the photon energy for the transition metal borides. Comparison of (**b**) the mean free path, (**d**) the half-value layer and (**f**) the tenth-value layer of selected transition metal borides with Pb.

In addition to the gamma shielding characteristics, the fast neutron removal cross-section parameter of the transition metal borides was also investigated to assess their total macroscopic cross-sections for fast neutrons. Note that boron is a great neutron absorber due to its high neutron cross-section, as seen in Figure 5b. Therefore, boron-containing materials are commonly used as neutron shields. The fast neutron removal cross-section values of the investigated transition metal borides vary in the range of 0.127–0.2 cm^{-1}, as seen in Figure 5a. Since the density of the compound and the weight fractions of the elements it contains are also important factors in calculating the fast neutron removal cross-section, the decreasing fast neutron removal cross-section order is ReB_2 > WB_2 > TaB_2 > HfB_2 > MoB_2 > NbB_2 > TiB_2 ZrB_2 > ErB_6 > DyB_6 > SmB_6 > NdB_6 > LaB_6 > EuB_6. Although the neutron capture cross-section of Sm is the highest, its contribution to the fast neutron removal cross-section is rather small. This is because of the relatively lower density of SmB_6 and weight fraction of Sm in SmB_6. It is also worth mentioning that several transition metal borides investigated in this study outperform many extensively used neutron shielding materials. The calculated fast neutron removal cross-section values of the ReB_2, WB_2, TaB_2, HfB_2, MoB_2, NbB_2, ZrB_2, and TiB_2 are significantly higher than those of lead (0.118 cm^{-1}), B_4C (0.141 cm^{-1}), NiO and PbO added borate glasses (0.111 cm^{-1}), concrete (0.094 cm^{-1}), graphite (0.077 cm^{-1}), and paraffin (0.077 cm^{-1}) [30,35,60].

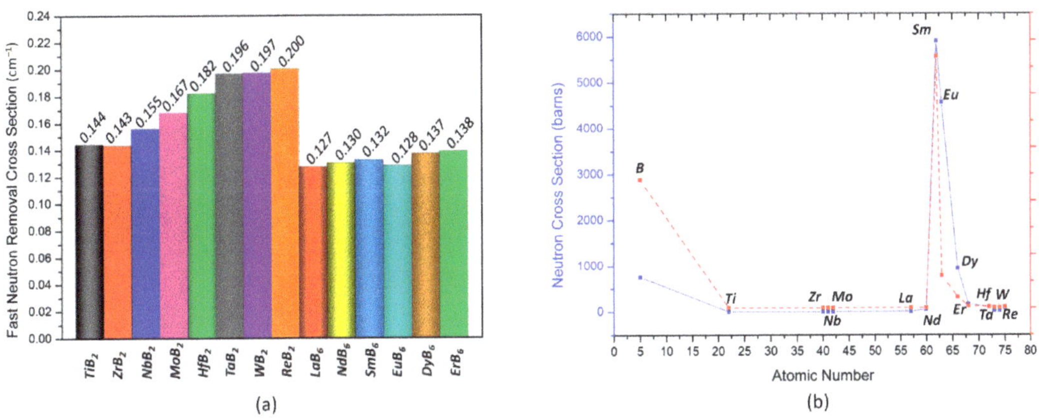

Figure 5. The fast neutron removal cross-sections (**a**) of the investigated transition metal borides and neutron capture cross-sections (**b**) of elements.

Further the fast neutron removal cross-sections of transition metal borides, the mass and linear attenuation factors for thermal (0.025 eV) and fast (4 MeV) neutrons at specific energies were also investigated (Figure 6). The results provided by NGCal software were found to be consistent with the neutron capture cross-section of elements that constitute the transition metal boride compounds shown in Figure 5b. SmB_6, EuB_6, and DyB_6 outperformed the rest of the transition metal borides on attenuating thermal neutrons with 0.025 eV energy, as seen in Figure 6a,c. On the other hand, for fast neutrons (4 MeV), DyB_6 showed the highest neutron attenuation Figure 6b,d. Due to dysprosium's larger neutron capture cross section at 4 MeV, DyB_6 even exceeded SmB_6.

The mass and linear neutron attenuation factors of widely used B_4C were also calculated to compare with selected transition metal borides (Table 2). For fast neutrons at 4 MeV energy, both the mass and linear neutron attenuation factors of B_4C were found to be lower than SmB_6 and DyB_6. The mass neutron attenuation factor of B_4C with 33.51172 at a neutron energy of 0.025 eV is slightly higher than SmB_6. Nevertheless, it decreases to 0.07678 for fast neutrons.

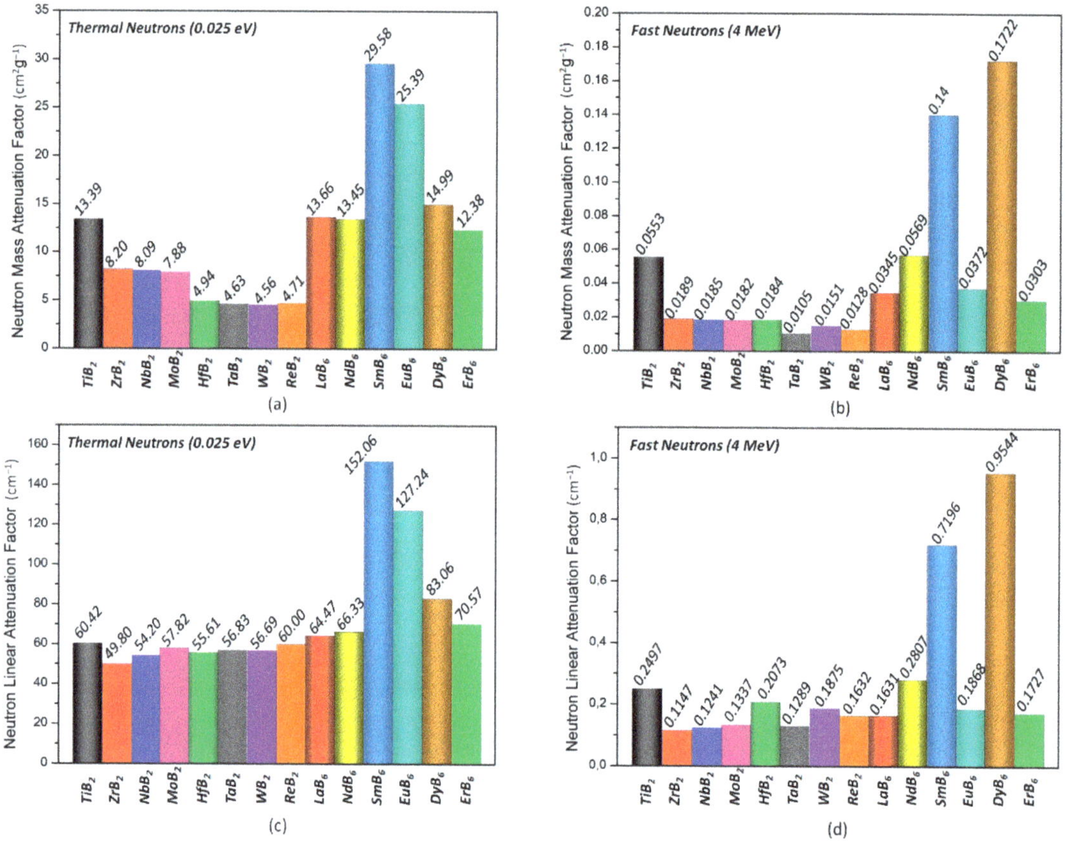

Figure 6. The mass (**a**,**b**) and linear (**c**,**d**) attenuation factors of selected compounds for thermal and fast neutrons.

Table 2. The comparison of the neutron attenuation factors of SmB_6 and DyB_6 with B_4C.

	The Mass Attenuation Factors		**The Linear Attenuation Factors**	
	Thermal Neutrons (0.025 eV)	**Fast Neutrons (4 MeV)**	**Thermal Neutrons (0.025 eV)**	**Fast Neutrons (4 MeV)**
B_4C	33.51172	0.07678	84.44954	0.19349
SmB_6	29.58439	0.14001	152.06374	0.71963
DyB_6	14.99297	0.17228	70.57155	0.17271

It can be concluded that transition metal borides provide excellent particle and photon shielding simultaneously, and they have the potential to surpass widely used radiation shielding materials Pb and B_4C. Nevertheless, further experimental tests are required to verify the theoretical model's accuracy and the transition metal borides' true potential.

4. Conclusions

In this study, photon and particle radiation shielding capacities of 14 different transition metal borides were evaluated, and the results were compared with commonly used radiation shielding materials. A thorough examination of the MAC, LAC, MFP, HVL, TVL, Z_{eff}, and FNRCS parameters demonstrated that transition metal borides possess excellent gamma and neutron radiation attenuation properties. MAC values of the transition metal borides exhibit a strong correlation with the atomic number of the transition metals and follow the order of ReB_2 > WB_2 > TaB_2 > HfB_2 > ErB_6 > DyB_6 > EuB_6 > SmB_6 > NdB_6 > LaB_6 > MoB_2 > NbB_2 > ZrB_2 > TiB_2. Meanwhile, MFP values of ReB_2, WB_2, and TaB_2 at 4 MeV are 1,99, 2.06, and 2.09, respectively. These results indicate that these three materials have superior gamma-ray attenuation capability than lead, which is the most commonly used radiation shielding material in nuclear applications.

On the other hand, the FNRCS ($\sum R$) of the transition metal borides depended on the density of the compounds and was found to increase as the density of the compounds increased. Seven transition metal borides, including ReB_2 (0.2 cm^{-1}), WB_2 (0.197 cm^{-1}), TaB_2 (0.196 cm^{-1}), HfB_2 (0.182 cm^{-1}), MoB_2 (0.167 cm^{-1}), NbB_2 (0.155 cm^{-1}), ZrB_2 (0.143 cm^{-1}), and TiB_2 (0.144 cm^{-1}), were found to have better fast neutron removal cross-sections than Pb (0.118 cm^{-1}), B_4C (0.141 cm^{-1}), borate glasses (0.111 cm^{-1}), concrete (0.094 cm^{-1}), and graphite (0.077 cm^{-1}). The ReB_2 has the highest fast neutron removal cross-section among the investigated transition metal borides. The transition metal boride's mass and linear neutron attenuation factors strongly correlate with the neutron capture cross-section of constituted elements. The borides of all rare-earth transition metals showed better neutron attenuation for thermal neutrons with 0.025 eV energy than the rest. For thermal neutrons, higher neutron capture cross-sections of Sm, Eu, and Dy led to the best neutron attenuation performance for SmB_6, EuB_6, and DyB_6 compounds. On the other hand, for fast neutrons with 4 MeV energy, DyB_6 outperformed. Most notably, SmB_6 and DyB_6 were found to have superior neutron attenuation factors than B_4C. The cost is also an important factor in designing sustainable radiation shielding materials. The low abundance of some rare earth elements and Re can increase the cost of their boride compounds. Nevertheless, the comparable price/performance ratio of WB_2, TaB_2, and SmB_6 with B_4C and nontoxicity over Pb make these compounds potential candidates for future developments in nuclear protection. Consequently, transition metal borides, especially ReB_2, WB_2, TaB_2, SmB_6, and DyB_6, show not only great gamma-ray shielding but also desirable neutron attenuation characteristics and can be excellent candidates for nuclear safety applications.

Author Contributions: C.A., conceptualization, investigation, data curation, visualization, writing—original draft, and writing—review, and editing; S.A., conceptualization, investigation, data curation, visualization, writing—original draft, and writing—review and editing. All authors have read and agreed to the published version of the manuscript.

Funding: This research received no external funding.

Informed Consent Statement: Not applicable.

Data Availability Statement: The data that support the findings of this study are available from the corresponding author upon reasonable request.

Acknowledgments: We acknowledge the support from the German Research Foundation and the Open Access Publication Fund of TU Berlin.

Conflicts of Interest: The authors declare no conflict of interest.

References

1. Ahmed, S.K.; Keole, S.R. Proton Therapy in the Adolescent and Young Adult Population. *Cancers* **2023**, *15*, 4269. [CrossRef] [PubMed]
2. Fernández-Arias, P.; Vergara, D.; Antón-Sancho, Á. Global Review of International Nuclear Waste Management. *Energies* **2023**, *16*, 6215. [CrossRef]
3. Moriarty, P.; Honnery, D. Review: The Energy Implications of Averting Climate Change Catastrophe. *Energies* **2023**, *16*, 6178. [CrossRef]
4. Sidhu, A.; Khan, N.; Phillips, C.; Briones, J.; Kapoor, A.; Zalewski, P.; Fleshner, N.E.; Chow, E.; Emmenegger, U. Prevalence and Prognostic Implications of PSA Flares during Radium-223 Treatment among Men with Metastatic Castration Resistant Prostate Cancer. *J. Clin. Med.* **2023**, *12*, 5604. [CrossRef] [PubMed]

5. Tuieng, R.J.; Cartmell, S.H.; Kirwan, C.C.; Sherratt, M.J. The Effects of Ionising and Non-Ionising Electromagnetic Radiation on Extracellular Matrix Proteins. *Cells* **2021**, *10*, 3041. [CrossRef]
6. Avcıoğlu, S.; Buldu, M.; Kaya, F.; Kaya, C. Chapter 20—Synthesis of nuclear-grade nano-sized boron carbide powders and its application in LDPE matrix composites for neutron shielding. In *Composite Materials: Manufacturing, Properties, and Applications*; Low, I.-M., Dong, Y., Eds.; Elsevier: Cambridge, MA, USA; Amsterdam, The Netherlands, 2021; pp. 543–579, ISBN 978-0-12-820512-9.
7. AVCIOĞLU, S. LDPE matrix composites reinforced with dysprosium-boron containing compounds for radiation shielding applications. *J. Alloys Compd.* **2022**, *927*, 166900. [CrossRef]
8. Sunitha, Y.; Kumar, S. 10B/11B isotopic ratio and atomic composition of boron carbide: Determination by proton induced γ-ray emission and proton elastic backscattering spectrometry. *Appl. Radiat. Isot.* **2017**, *128*, 28–35. [CrossRef]
9. UZUN, S. Investigation of boron element and its use in nuclear reactors. *J. Boron* **2023**, *8*, 25–30. [CrossRef]
10. Pitto-Barry, A. Polymers and boron neutron capture therapy (BNCT): A potent combination. *Polym. Chem.* **2021**, *12*, 2035–2044. [CrossRef]
11. Monti Hughes, A.; Hu, N. Optimizing Boron Neutron Capture Therapy (BNCT) to Treat Cancer: An Updated Review on the Latest Developments on Boron Compounds and Strategies. *Cancers* **2023**, *15*, 4091. [CrossRef]
12. Avcıoğlu, S.; Buldu, M.; Kaya, F.; Üstündağ, C.B.; Kam, E.; Menceloğlu, Y.Z.; Kaptan, H.Y.; Kaya, C. Processing and properties of boron carbide (B4C) reinforced LDPE composites for radiation shielding. *Ceram. Int.* **2020**, *46*, 343–352. [CrossRef]
13. Özcan, M.; Avcıoğlu, S.; Kaya, C.; Kaya, F. Boron carbide reinforced electrospun nanocomposite fiber mats for radiation shielding. *Polym. Compos.* **2023**, *44*, 4155–4167. [CrossRef]
14. Özcan, M.; Kam, E.; Kaya, C.; Kaya, F. Boron-containing nonwoven polymeric nanofiber mats as neutron shields in compact nuclear fusion reactors. *Int. J. Energy Res.* **2022**, *46*, 7441–7450. [CrossRef]
15. More, C.V.; Alsayed, Z.; Badawi, M.S.; Thabet, A.A.; Pawar, P.P. Polymeric composite materials for radiation shielding: A review. *Environ. Chem. Lett.* **2021**, *19*, 2057–2090. [CrossRef]
16. Nambiar, S.; Yeow, J.T.W. Polymer-composite materials for radiation protection. *ACS Appl. Mater. Interfaces* **2012**, *4*, 5717–5726. [CrossRef] [PubMed]
17. Almurayshid, M.; Alsagabi, S.; Alssalim, Y.; Alotaibi, Z.; Almsalam, R. Feasibility of polymer-based composite materials as radiation shield. *Radiat. Phys. Chem.* **2021**, *183*, 109425. [CrossRef]
18. Okonkwo, U.C.; Idumah, C.I.; Okafor, C.E.; Ezeani, E.O. Construction of radiation attenuating polymeric nanocomposites and multifaceted applications: A review. *Polym. Plast. Technol. Mater.* **2023**, *62*, 1639–1661. [CrossRef]
19. Qi, Z.; Yang, Z.; Li, J.; Guo, Y.; Yang, G.; Yu, Y.; Zhang, J. The Advancement of Neutron-Shielding Materials for the Transportation and Storage of Spent Nuclear Fuel. *Materials* **2022**, *15*, 3255. [CrossRef]
20. Yasin, T.; Khan, M.N. High density polyethylene/boron carbide composites for neutron shielding. *e-Polymers* **2008**, *8*, 670–676. [CrossRef]
21. Salimi, M.; Ghal-Eh, N.; Amirabadi, E.A. Characterization of a new shielding rubber for use in neutron–gamma mixed fields. *Nucl. Sci. Tech.* **2018**, *29*, 36. [CrossRef]
22. Abdel-Aziz, M.M.; Gwaily, S.E.; Makarious, A.S.; El-Sayed Abdo, A. Ethylene-propylene diene rubber/low density polyethylene/boron carbide composites as neutron shields. *Polym. Degrad. Stab.* **1995**, *50*, 235–240. [CrossRef]
23. Avcıoğlu, C.; Bekheet, M.F.; Artır, R. Radiation shielding and mechanical properties of mullite-zirconia composites fabricated from investment-casting shell waste. *J. Mater. Res. Technol.* **2023**, *24*, 5883–5895. [CrossRef]
24. Kassem, S.M.; Abdel Maksoud, M.I.A.; Ghobashy, M.M.; El Sayed, A.M.; Ebraheem, S.; Helal, A.I.; Ebaid, Y.Y. Novel flexible and lead-free gamma radiation shielding nanocomposites based on LDPE/SBR blend and $BaWO_4/B_2O_3$ heterostructures. *Radiat. Phys. Chem.* **2023**, *209*, 110953. [CrossRef]
25. Kaur, T.; Sharma, J.; Singh, T. Review on scope of metallic alloys in gamma rays shield designing. *Prog. Nucl. Energy* **2019**, *113*, 95–113. [CrossRef]
26. Kharita, M.H.; Yousef, S.; AlNassar, M. Review on the addition of boron compounds to radiation shielding concrete. *Prog. Nucl. Energy* **2011**, *53*, 207–211. [CrossRef]
27. Sallam, O.I.; Madbouly, A.M.; Elalaily, N.A.; Ezz-Eldin, F.M. Physical properties and radiation shielding parameters of bismuth borate glasses doped transition metals. *J. Alloys Compd.* **2020**, *843*, 156056. [CrossRef]
28. Sallam, O.I.; Madbouly, A.M.; Ezz-Eldin, F.M. Impact of Nd^{3+} additive on the radiation shielding competence of borosilicate glasses fabricated from agro-waste materials. *J. Non-Cryst. Solids* **2022**, *590*, 121691. [CrossRef]
29. Kavaz, E.; Ersundu, M.Ç.; Ersundu, A.E.; Tekin, H.O. Synthesis and characterization of newly developed phosphate-based glasses through experimental gamma-ray and neutron spectroscopy methods: Transmission and dose rates. *Ceram. Int.* **2022**, *48*, 13842–13849. [CrossRef]
30. Ilik, E.; Kavaz, E.; Kilic, G.; Issa, S.A.M.; Zakaly, H.M.H.; Tekin, H.O. A closer-look on Copper(II) oxide reinforced Calcium-Borate glasses: Fabrication and multiple experimental assessment on optical, structural, physical, and experimental neutron/gamma shielding properties. *Ceram. Int.* **2022**, *48*, 6780–6791. [CrossRef]
31. Al-Ghamdi, H.; Almuqrin, A.H.; Koubisy, M.S.I.; Mahmoud, K.A.; Sayyed, M.I.; Darwish, M.A.; Henaish, A.M.A. Zinc-lead-borate glasses doped with dysprosium oxide: Structure, optical, and radiation shielding features. *Optik* **2021**, *246*, 167765. [CrossRef]

32. Kavaz, E.; Ilik, E.; Kilic, G.; ALMisned, G.; Tekin, H.O. Synthesis and experimental characterization on fast neutron and gamma-ray attenuation properties of high-dense and transparent Cadmium oxide (CdO) glasses for shielding purposes. *Ceram. Int.* **2022**, *48*, 23444–23451. [CrossRef]
33. Lakshminarayana, G.; Elmahroug, Y.; Kumar, A.; Tekin, H.O.; Rekik, N.; Dong, M.; Lee, D.-E.; Yoon, J.; Park, T. Detailed Inspection of γ-ray, Fast and Thermal Neutrons Shielding Competence of Calcium Oxide or Strontium Oxide Comprising Bismuth Borate Glasses. *Materials* **2021**, *14*, 2265. [CrossRef] [PubMed]
34. ALMisned, G.; Tekin, H.O.; Kavaz, E.; Bilal, G.; Issa, S.A.M.; Zakaly, H.M.H.; Ene, A. Gamma, Fast Neutron, Proton, and Alpha Shielding Properties of Borate Glasses: A Closer Look on Lead (II) Oxide and Bismuth (III) Oxide Reinforcement. *Appl. Sci.* **2021**, *11*, 6837. [CrossRef]
35. Sallam, O.I.; Madbouly, A.M.; Moussa, N.L.; Abdel-Galil, A. Impact of radiation on CoO-doped borate glass: Lead-free radiation shielding. *Appl. Phys. A* **2022**, *128*, 70. [CrossRef]
36. Sallam, O.I.; Issa, S.A.M.; Rashad, M.; Madbouly, A.M.; Tekin, H.O.; Badawi, A.; Hamdy, A.; Zakaly, H.M.H. Impact of molybdenum on optical, structure properties and gamma radiation shielding parameters of bor-ophosphate glass: Intensive experiment investigations. *Radiat. Phys. Chem.* **2022**, *198*, 110140. [CrossRef]
37. Madbouly, A.M.; Sallam, O.I.; Issa, S.A.M.; Rashad, M.; Hamdy, A.; Tekin, H.O.; Zakaly, H.M.H. Experimental and FLUKA evaluation on structure and optical properties and γ-radiation shielding capacity of bismuth borophosphate glasses. *Prog. Nucl. Energy* **2022**, *148*, 104219. [CrossRef]
38. Xiao, M.; Qin, Q.; He, X.; Li, F.; Wang, X. Shielding Capability Research on Composite Base Materials in Hybrid Neutron-Gamma Mixed Radiation Fields. *Materials* **2023**, *16*, 2084. [CrossRef]
39. Arif Sazali, M.; Alang Md Rashid, N.K.; Hamzah, K. A review on multilayer radiation shielding. *IOP Conf. Ser. Mater. Sci. Eng.* **2019**, *555*, 12008. [CrossRef]
40. Daneshvar, H.; Milan, K.G.; Sadr, A.; Sedighy, S.H.; Malekie, S.; Mosayebi, A. Multilayer radiation shield for satellite electronic components protection. *Sci. Rep.* **2021**, *11*, 20657. [CrossRef]
41. Pangilinan, L.E.; Hu, S.; Hamilton, S.G.; Tolbert, S.H.; Kaner, R.B. Hardening Effects in Superhard Transition-Metal Borides. *Acc. Mater. Res.* **2022**, *3*, 100–109. [CrossRef]
42. Zoli, L.; Galizia, P.; Silvestroni, L.; Sciti, D. Synthesis of group IV and V metal diboride nanocrystals via borothermal reduction with sodium borohydride. *J. Am. Ceram. Soc.* **2018**, *101*, 2627–2637. [CrossRef]
43. MERTDİNÇ, S.; AĞAOĞULLARI, D.; ÖVEÇOĞLU, M.L. TaB/TaB_2 Tozlarının Mekanokimyasal Sentezleme ile Üretimi ve Karakterizasyonu. *AKÜ FEMÜBİD* **2019**, *19*, 465–473.
44. Yan, G.; Feng, Y.; Fu, B.; Liu, C.; Ji, P.; Zhang, P.; Zhou, L. Supercondcuting properties in MgB_2/Fe wires prepared by PIT method. *Chin. Sci. Bull.* **2003**, *48*, 1331–1333. [CrossRef]
45. Materials Explorer. *Materials Data on LaB6 by Materials Project*; Materials Explorer: Riverside, CA, USA, 2022.
46. Materials Project App. *Materials Data on LuB6 (SG:221) by Materials Project*; Materials Explorer: Riverside, CA, USA, 2014.
47. Cahill, J.T.; Graeve, O.A. Hexaborides: A review of structure, synthesis and processing. *J. Mater. Res. Technol.* **2019**, *8*, 6321–6335. [CrossRef]
48. Jain, A.; Ong, S.P.; Hautier, G.; Chen, W.; Richards, W.D.; Dacek, S.; Cholia, S.; Gunter, D.; Skinner, D.; Ceder, G.; et al. Commentary: The Materials Project: A materials genome approach to accelerating materials innovation. *APL Mater.* **2013**, *1*, 011002. [CrossRef]
49. Şakar, E.; Özpolat, Ö.F.; Alım, B.; Sayyed, M.I.; Kurudirek, M. Phy-X/PSD: Development of a user friendly online software for calculation of parameters relevant to radiation shielding and dosimetry. *Radiat. Phys. Chem.* **2020**, *166*, 108496. [CrossRef]
50. Sakar, E.; Guler, O.; Alım, B.; Say, Y.; Dikici, B. A comprehensive study on structural properties, photon and particle attenuation competence of CoNiFeCr-Ti/Al high entropy alloys (HEAs). *J. Alloys Compd.* **2023**, *931*, 167561. [CrossRef]
51. Şakar, E.; Alım, B.; Ertap, H.; Karabulut, M. An extensive survey on radiation protection features of novel hafnium iron-borophosphate glasses: Experimental and theoretical study. *Prog. Nucl. Energy* **2023**, *160*, 104690. [CrossRef]
52. Gökçe, H.S.; Güngör, O.; Yılmaz, H. An online software to simulate the shielding properties of materials for neutrons and photons: NGCal. *Radiat. Phys. Chem.* **2021**, *185*, 109519. [CrossRef]
53. Hila, F.C.; Asuncion-Astronomo, A.; Dingle, C.A.M.; Jecong, J.F.M.; Javier-Hila, A.M.V.; Gili, M.B.Z.; Balderas, C.V.; Lopez, G.E.P.; Guillermo, N.R.D.; Amorsolo, A.V. EpiXS: A Windows-based program for photon attenuation, dosimetry and shielding based on EPICS2017 (ENDF/B-VIII) and EPDL97 (ENDF/B-VI.8). *Radiat. Phys. Chem.* **2021**, *182*, 109331. [CrossRef]
54. Zhang, T.; Li, Y.; Yuan, Y.; Cui, K.; Wei, W.; Wu, J.; Qin, W.; Wu, X. Spatially confined Bi2O3–Ti3C2T hybrids reinforced epoxy composites for gamma radiation shielding. *Compos. Commun.* **2022**, *34*, 101252. [CrossRef]
55. Hubbell, J.H. *Photon Cross Sections, Attenuation Coefficients, and Energy Absorption Coefficients from 10 keV to 100 GeV*; U.S. Department of Commerce, The National Standard Reference Data System: Washington, DC, USA, 1969.
56. Eisa, M.H.; Shen, H.; Yao, H.Y.; Mi, Y.; Zhou, Z.Y.; Hu, T.D.; Xie, Y.N. Studies on absorption coefficient near edge of multi elements. *J. Quant. Spectrosc. Radiat. Transf.* **2005**, *96*, 503–511. [CrossRef]
57. Cremer, J.T. Chapter 4—X-ray Optics. In *Advances in Imaging and Electron Physics*; Cremer, J.T., Ed.; Elsevier: Amsterdam, The Netherlands, 2012; pp. 497–559. ISBN 1076-5670.
58. Zegaoui, A.; Derradji, M.; Ghouti, H.A.; Medjahed, A.; Zu, L.-W.; Liu, W.-B.; Wang, J.; Dayo, A.Q.; Liu, Y.-G. Synergetic effects of short carbon/basalt hybrid fibers on the mechanical, thermal and nuclear shielding properties of DCBA/BA-a resin composites. *Compos. Commun.* **2019**, *15*, 179–185. [CrossRef]

59. Crompton, T.R. Chapter 4—Metals in Surface, Ground, and Mineral Waters. In *Determination of Metals in Natural Waters, Sediments and Soils*; Crompton, T.R., Ed.; Elsevier: Amsterdam, The Netherlands; Boston, MA, USA, 2015; pp. 161–178, ISBN 978-0-12-802654-0.
60. Singh, V.P.; Badiger, N.M. Investigation of Gamma and Neutron Shielding Parameters for Borate Glasses Containing NiO and PbO. *Phys. Res. Int.* **2014**, *2014*, 954958. [CrossRef]

Disclaimer/Publisher's Note: The statements, opinions and data contained in all publications are solely those of the individual author(s) and contributor(s) and not of MDPI and/or the editor(s). MDPI and/or the editor(s) disclaim responsibility for any injury to people or property resulting from any ideas, methods, instructions or products referred to in the content.

materials

MDPI

Article

Inside the Borate Anomaly: Leveraging a Predictive Modelling Approach to Navigate Complex Composition–Structure–Property Relationships in Oxyhalide Borate Glasses

Brenna Kettlewell [1] and Daniel Boyd [2,*]

[1] School of Biomedical Engineering, Dalhousie University, Halifax, NS B3H 4R2, Canada; brenna.kettlewell@dal.ca

[2] Department of Applied Oral Sciences, Faculty of Dentistry, Dalhousie University, 5981 University Avenue, P.O. Box 15000, Halifax, NS B3H 4R2, Canada

* Correspondence: d.boyd@dal.ca

Abstract: This study employs a systematic and predictive modelling approach to investigate the structure and properties of multi-component borate glasses. In particular, this work is focused on understanding the individual and interaction effects of multiple constituents on several material properties. By leveraging advanced modeling techniques, this work examines how the inclusion and variation of B_2O_3, CaF_2, TiO_2, ZnO, and Na_2CO_3 influence the glass network, with particular attention to modifier fractions $\geq$ 30 mol%. This research addresses the gap in knowledge regarding the complex behavior of borate glasses in this high modifier fraction range, known as the borate anomaly, where prediction of glass structure and properties becomes particularly challenging. The use of a design of mixtures (DoM) approach facilitated the generation of polynomial equations indicating the influence of mixture components on various responses, enabling the prediction and optimization of glass properties over broad compositional ranges despite being within the anomalous region. This methodical approach not only advances our understanding of borate glass systems but also underscores the importance of predictive modelling in the accelerated design and development of novel glass materials for diverse applications.

Keywords: borate glass; design of mixtures; borate anomaly; optimization; predictive modelling

1. Introduction

Relative to traditional tissue engineering constructs, bioactive glasses are structurally simple, yet functionally complex materials capable of promoting a range of desirable material and host responses via the controlled release of multiple therapeutic inorganic ions (TII) [1,2]. The localized delivery of these bioactive ions offers significant advantages over other chemical and biochemical compounds (e.g., carbohydrates, lipids, proteins, and nucleic acids) for drug delivery. These advantages include, but are not limited to, (i) high thermal stability and thus easier manufacturing, processing, and storage, (ii) the ability to interact synergistically with other ions, (iii) reduced cost, and (iv) the elimination of additional drug loading steps as the therapeutic ions are directly incorporated into the synthesis of the glass itself [3,4]. Similar to typical organic drug biomolecules, such as growth factors, TIIs have been demonstrated to alter cellular functions and metabolism as enzyme cofactors and/or by activating ion channels or secondary signaling [3,5]. The controlled release of TIIs is dependent on the composition of the glass and how each constituent alters the structural characteristics of the network. Therefore, to understand how TIIs can be harnessed to modulate various biological responses (e.g., antibacterial, anti-inflammatory, osteoinductive, and angiogenic processes), one must first understand the physical, chemical, and structural properties of a given glass system.

Within the vast landscape of glass compositions capable of eliciting desired material and host responses, it is noteworthy that a significant focus in the investigation of bioactive glasses centers around silicate glass networks [6]. Despite their prominence, it is important to acknowledge that these glass systems present challenges in some indications, such as slow and incomplete degradation. This is attributed to the low solubility of silica and the formation of a silica-rich gel layer on the glass surface, which, in turn, restricts the diffusion of ions into the solution [7–9]. In the context of bone tissue engineering, for example, this translates into slow and incomplete conversion to hydroxyapatite (HA), which results in decreased rates of bone formation, longer healing times, and residual materials at the implant site [9,10]. As a result of such limitations, there exists opportunities to discover and explore a broader spectrum of glass compositions with distinct degradation profiles and payloads of TIIs. Contrary to their silicate counterparts, and of increasing interest in the literature, borate glasses undergo complete degradation, a process that can be adjusted through the selection of specific TIIs, characterized as either network modifying, network forming, or intermediate elements, and their inclusion into the molecular network of the glass [11]. Consequently, borate glasses are gaining increasing interest in biomedical applications for their ability to act as fully resorbable carrier systems for the controlled and localized delivery of TIIs [12–14].

Unlike the properties of silicate glasses, which vary monotonically with the network-modifying fraction of oxides contained within the glass structure, borate glass properties exhibit extrema in response to specific quantities of network modifier(s) and the presence of certain basic structural and superstructural units [11]. For instance, upon the addition of a network modifier to the borate glass network, the trigonal $BØ_3$ structural units convert to tetrahedral $BØ_4^-$ units. This conversion takes place until an added modifier fraction of approximately 30 mol% has been reached, whereafter, further addition of modifier leads to the formation of non-bridging oxygens and the subsequent decrease in $BØ_4^-$ units [11,15]. An increased number of four-coordinated boron structural units increases the connectivity of the glass network, thus providing enhanced hydrolytic stability and resistance to degradation. In contrast, the formation of non-bridging oxygens leads to a more disrupted glass network and, hence, a more reactive material. Therefore, the type and fraction of network-modifying elements included in the network can be tailored to modulate the structure and dissolution of borate glasses, providing improved control and predictability over a range of material and host responses.

The non-linear structural behavior exhibited by borate glasses is referred to as the borate anomaly [11,16]. This behavior is the basis upon which glasses containing a network modifier fraction greater than 30 mol% are notoriously difficult to predict and thus vastly understudied [6,11]. Furthermore, the behavior of borate glass systems has primarily been reported for binary glass substitutions of alkali or alkaline earth network modifiers [11]. Hence, the properties of ternary and quaternary borate glass systems are more poorly understood, while those of greater complexity, comprising multiple cations and anions with variable valences and cationic field strengths, are even less predictable.

As such, the ongoing investigation of borate glass systems is currently faced with exciting opportunities for growth. The potential for streamlined development of these materials, especially those containing more than two components, substitutions from groups beyond alkali and alkaline earth metals, and modifier fractions exceeding 30 mol%, is substantial. However, as with other advanced material technologies, the accelerated translation of these materials towards clinical use has been somewhat hindered by a current lack of comprehensive knowledge surrounding their structure and properties [17–19]. This limited understanding is associated with the use of traditional trial-and-error style approaches and/or one-variable-at-a-time (OVAT) methodologies in the study of glass materials. While these approaches have played a significant role in the design and development of novel glass materials, the sequential adjustments of individual variables characteristic of these methodologies can be extremely time-consuming, resource intensive, and impractical for investigating extensive combinatorial palettes of glass chemistries. Specifically, trial-and-error or OVAT approaches may inadvertently overlook or underestimate the complexity of multi-component glasses and the likelihood for elements to interact with each other, forming synergistic relationships within the network that may influence material and host responses. Hence, this complicates our ability to predict the individual *and* interaction effects of glass constituents, especially in systems comprising multiple cations and anions with variable valences and cationic field strengths. Without the ability to gain insights into these complex relationships, traditional methodologies may not efficiently navigate the multidimensional space of compositional factors, thus precluding the simultaneous optimization of multiple variables. In light of these considerations, there is a clear imperative for more sophisticated and comprehensive approaches to ensure scientific rigor, accuracy, and sustainability in the design of glass materials [20–22].

Further investigation of borate glass materials is necessary to support their accelerated discovery, design, and deployment across a wide range of indications, including biomedical, nuclear, and industrial applications. Given the complex nature of these systems, quantifying and predicting the composition–structure–property relationships within these materials requires comprehensive and systematic evaluation. In this regard, the Materials Genome Initiative (MGI) advanced a new paradigm for accelerated materials innovation [19]. The approach of the MGI encourages the combination of experimentation, theory, and computation to permit the high-throughput screening, prediction, and optimization of new materials and promotes the collaboration of researchers based on shared access to numerous materials datasets [19,23]. One systematic approach that conforms to the objectives of the MGI and supports the accurate prediction of materials properties across broad compositional ranges is the design of mixtures (DoM) approach. This approach simultaneously integrates experimental methods and advanced modelling to generate statistically driven polynomial equations that indicate the relative influences of mixture components, both individually and interactively, on a given response. Using surface regression methodologies, this approach permits the prediction and optimization of materials for a wide variety of desirable properties, which, unlike traditional trial-and-error style approaches, elucidates a more deliberate and streamlined design of materials [24–28]. Accordingly, the primary objective of the current study is to employ a DoM approach to predict the individual and interaction effects of multiple constituents on the physical and chemical properties of multi-component, degradable borate glasses with modifier fractions $\geq$ 30 mol%, a compositional region that is poorly understood and difficult to predict.

2. Material and Methods

2.1. Material Design and Synthesis

The compositions for each glass were established using Design-Expert Software (Version 13). A DoM I-optimal quadratic model was utilized based on five components (B_2O_3, CaF_2, TiO_2, ZnO, Na_2CO_3). Specifically, the I-optimal algorithm was chosen to accommodate this complex mixture, comprising multiple components with varying compositional ranges. Furthermore, this algorithm chooses formulations (i.e., runs) that minimize the integral of prediction variance across the design space, making it the recommended algorithm for optimization of the materials. A quadratic model was necessary to gain insight into not only the individual effects of each component (as provided with a linear model), but also the second-order interaction effects. Various test melts were conducted to confirm the compositional constraints of each constituent (Table 1). Based on these constraints, the model yielded a total of 23 randomized glass formulations for investigation (Table 2), consisting of 15 model points, 4 lack of fit points, and 4 replicates. Replicate compositions were included to verify reproducibility in glass synthesis and characterization processes and to increase the predictive power of the model. Using standard analysis of variance (ANOVA), a quadratic Scheffe polynomial equation was fitted to each measured response based on the following equation:

$$Y = \beta_1 A + \beta_2 B + \beta_3 C + \beta_4 D + \beta_5 E + \beta_{12} AB + \beta_{13} AC + \beta_{14} AD + \beta_{15} AE + \beta_{23} BC + \beta_{24} BD + \beta_{25} BE + \beta_{34} CD + \beta_{35} CE + \beta_{45} DE$$

where A to E represent the compositional factors, $\beta_{1\text{–}5}$ coefficients represent the effect of the individual compositional factors A to E, and $\beta_{12\text{–}45}$ are the coefficients of regression, which represent the effects of the interactions between the compositional factors A to E [29]. This response surface regression model approach was used to quantitatively determine the relative impacts of each glass constituent, both individually and interactively, on variable material properties, such as density, T_g, and fraction of B3 and B4 structural units.

Table 1. Mixture components and design constraint summary.

Component	Units	Minimum [mol%]	Maximum [mol%]	Coded Low	Coded High	Mean	Std. Dev.
B_2O_3	Mol%	45	70	+0 ↔ 45	+0.55 ↔ 70	59.69	7.65
CaF_2	Mol%	0	20	+0 ↔ 0	+0.44 ↔ 20	11.70	7.08
TiO_2	Mol%	0	10	+0 ↔ 0	+0.22 ↔ 10	5.32	3.90
ZnO	Mol%	5	20	+0 ↔ 5	+0.33 ↔ 20	13.28	5.78
Na_2CO_3	Mol%	5	15	+0 ↔ 5	+0.22 ↔ 15	10.01	4.06

Table 2. Compositions of the 23 glasses (mol%), as identified by the constraints outlined in Table 1. Within the table there are four replicate compositions consisting of 1 and 12, 3 and 6, 4 and 5, and 17 and 21.

	B_2O_3	CaF_2	TiO_2	ZnO	Na_2CO_3
1	70.00	13.47	3.10	5.00	8.43
2	45.00	20.00	10.00	20.00	5.00
3	54.61	11.44	10.00	13.33	10.62
4	63.26	2.08	4.76	20.00	9.90
5	63.26	2.08	4.76	20.00	9.90
6	54.61	11.44	10.00	13.33	10.62
7	64.78	6.38	8.83	5.00	15.00
8	64.75	8.65	6.16	15.43	5.00
9	45.00	20.00	7.77	12.23	15.00
10	70.00	0.00	10.00	12.75	7.25
11	55.24	13.99	5.78	19.99	5.00
12	70.00	13.47	3.10	5.00	8.43
13	50.44	14.34	0.22	20.00	15.00
14	68.01	3.17	1.36	12.46	15.00
15	60.00	20.00	0.00	5.00	15.00
16	53.20	17.96	9.74	5.00	14.10
17	56.49	20.00	2.01	13.54	7.97
18	60.00	20.00	10.00	5.00	5.00
19	55.00	0.00	10.00	20.00	15.00
20	69.82	6.73	0.00	18.44	5.00
21	56.49	20.00	2.01	13.54	7.97
22	65.89	16.23	0.00	12.88	5.00
23	57.01	7.56	2.86	17.57	15.00

Glasses were synthesized via the traditional melt quenching technique. Succinctly, analytical-grade boric anhydride, calcium fluoride, titanium (IV) oxide, zinc oxide, and sodium carbonate reagents (Sigma-Aldrich, Burlington, MA, USA) were weighed out in accordance with compositions listed in Table 2 and homogenized for 1 h prior to being transferred to 50 mL Pt-Rh crucibles (XRF Scientific, Montreal, QC, Canada) for melting [30]. The crucible was placed in a high-temperature box furnace (Carbolite, RHF 16/3) and followed a two-step thermal process; materials were heated from room temperature to 850 °C (10 °C/min) and held for 2 h for calcination, then ramped from 850 °C to 1200 °C

(10 °C/min) and held for 1 h for melting. Each melt was quenched using stainless steel plates, and the resulting frit was ground and sieved (ASTM E-11 [31] compliant sieves) to obtain particles in size ranges of <90 μm, 90–710 μm, and > 710 μm diameter (Superla Sieve™, Newark Wire Cloth Company, Clifton, NJ, USA). Glasses were stored in desiccated conditions until the time of subsequent analysis.

2.2. Particle Size Analysis

A Malvern Mastersizer 3000 laser diffraction particle size analyzer was used as per the manufacturer's instructions. Samples of each glass, sieved to 90–710 μm, were suspended in deionized water to obtain an obscuration value for the suspension between 5 and 8%. The glass suspension was then measured using both a blue (λ = 470 nm) and a red (λ = 632.8 nm) laser. Each glass suspension was measured five times (n = 5), and particle size distribution data were reported as the mean diameter Dx90, Dx50, and Dx10 (particle diameters at 90%, 50%, and 10% cumulative size, respectively).

2.3. X-ray Diffraction

X-ray diffraction analysis was performed using a Bruker D2 phaser (Bruker, Billerica, MA, USA) equipped with a Lynx-Eye 1D detector (linear array). Powder specimens of each composition were loaded into a PMMA sample holder and analyzed in the region between $10° \leq 2\theta \leq 60°$ with a step size of $2\theta = 0.03$ and a step time of 2 s. XRD spectra were analyzed using Bruker Diffrac.Eva (https://www.bruker.com/en/products-and-solutions/diffractometers-and-x-ray-microscopes/x-ray-diffractometers/diffrac-suite-software/diffrac-eva.html) to calculate the relative fraction of crystalline material in each sample by evaluating the amorphous baseline to crystalline peak separation. Where applicable, Diffrac.Eva software was used to identify crystalline phases through peak matching against reference databases.

2.4. Density

An AccuPyc 1340 helium pycnometer (Micromeritics, Norcross, GA, USA) equipped with a 1 cm^3 insert chamber was used to determine the density of each glass composition as per manufacturer instructions [32]. Briefly, 0.5–0.6 g of glass powder was used for each measurement, and each measurement consisted of 10 fill and purge cycles (run in triplicate). The results were reported as the average ± standard deviation (SD) of three replicate measurements.

2.5. Differential Scanning Calorimetry

A simultaneous thermal analysis Luxx 409 PC STA with Auto-Sampler (Netzsch-Geratebau-GMBH, Exton, PA, USA) was used to analyze each glass composition. Approximately 20–30 mg of glass powder was weighed out, placed into Pt-Rh crucibles, and heated at 10 °C/min from 30 to 1000 °C (Standard Test Method for Assignment of the Glass Transition Temperatures by Differential Scanning Calorimetry, ASTM E1356 [33]). Proteus Thermal Analysis software (Version 8.0.2) was used to determine the extrapolated glass transition temperature (T_g) (inflection).

2.6. ^{11}B MAS NMR

^{11}B magic angle spinning (MAS) NMR spectra were acquired using a 16.4 T Bruker Avance NMR spectrometer (^{11}B Larmor frequency = 224.67 MHz) equipped with a 2.5 mm HX probe head operating in single resonance mode. ^{11}B parameters were calibrated on solid $NaBH_4$, which was also used as a secondary chemical shift reference (−42.1 ppm relative to $BF_3 \cdot Et_2O$). All samples were spun at 20 kHz MAS frequency, producing 64 scans. A single 0.63 μs pulse was used for all experiments, corresponding to a pulse angle of roughly 15° in a nearby cubic environment of $NaBH_4$. Spin lattice relaxation times were determined separately for the B3 and B4 groups using a saturation recovery sequence and were found to be around 6–9 s. The pulse repetition times were chosen to be 5 s. The boron background was removed by subtracting the spectrum of the used empty rotor from the sample spectra. After baseline correction with a spline function, the peak heights, widths, and integral values were determined for the B3 and B4 groups using XWinNMR's native tools.

2.7. Optimization and Validation

A response surface optimization study using Design-Expert Software (Version 13) was conducted to predict two new desirable glass compositions tailored to meet a set of optimization criteria (listed in Table 3). The optimization criteria for the two formulations (denoted as optimized glass A and B) were chosen to yield solutions with distinct compositions. This design permits a comprehensive investigation of model predictability across two glass systems exhibiting variable structure and properties. For instance, it was preferred that optimized glass A feature a B_2O_3 content at the upper end of the compositional constraints (70 mol%), thus representing a composition on the upper limit of the anomalous region (i.e., modifier fraction = 30 mol%), as defined by the binary borate glass literature [11,15]. In contrast, optimized glass B was preferred to have a B_2O_3 content within the specified compositional range (45 to 70 mol%), thus representing a composition within the anomalous region. Design-Expert software systematically generated numerous potential solutions based on the optimization criteria selected.

Table 3. Criteria for optimized glass formulations.

Optimization Criteria						
Factor/Response	**Units**	**Lower Limit**	**Upper Limit**	**Importance**	**Goal: Optimized Glass A**	**Goal: Optimized Glass B**
B_2O_3	Mol%	45	70	3	Maximize	In range
CaF_2	Mol%	0	20	3	In range	In range
TiO_2	Mol%	0	10	3	In range	In range
ZnO	Mol%	5	20	3	In range	In range
Na_2CO_3	Mol%	5	15	3	In range	In range
Density	g/cm^3	2.4	3.1	5	Target = 2.5	Target = 2.9
%B4	%	23.59	42.41	3	Minimize	Maximize
T_g	°C	471.6	563.8	3	Target = 540	Target = 510
Crystallinity	%	0.86	19.62	3	Minimize	Minimize

The two optimal glass formulations were chosen based on the desirability score assigned to each solution. The desirability score serves as an objective function, ranging from zero (outside of the limits) to one (at the goal), and is contingent on how closely the lower and upper limits align with the actual optimum. The numerical optimization process identifies a point that maximizes the desirability function. Therefore, at both optimized settings, the solution with the highest desirability score was selected as the preferred candidate.

The two chosen optimized formulations were synthesized using the same parameters as the initial 23 glasses. Subsequently, density, DSC, and ^{11}B MAS NMR analyses were repeated on the optimized glasses using the same parameters as the initial 23 glasses. Furthermore, XRD analysis was conducted on the optimized glasses to verify the absence of identifiable crystalline species, with a doubled step time to account for the use of a low background sample holder. The actual experimental values were compared to the predicted values of the model using a two-sided confidence interval of 95% to validate the predictive power of the model. If actual values fell within the 95% prediction intervals, this indicated that the model predicted accurately at those optimized settings.

3. Results

3.1. Glass Synthesis

All compositions were successfully synthesized and formed flowable melts capable of being poured and quenched. The quenched glasses varied in opacity. Compositions #7, 10, and 19 were almost entirely white opaque materials upon quenching, which were later verified as partially crystalline materials via XRD analysis, exhibiting % crystallinities of 12%, 17%, and 15%, respectively.

3.2. Particle Size Analysis

Each glass was processed to meet the particle size distribution of 90–710 μm. The Dx10, Dx50, and Dx90 values for each glass composition were measured (Table 4) to confirm repeatability and reproducibility in the glass processing.

Table 4. Material characterization results for particle size, density, crystallinity, glass transition temperature, and boron structural units for each glass composition.

	Dx(10) [μm]	Dx(50) [μm]	Dx(90) [μm]	Density [g/cm^3]	% Crystallinity [%]	T_g [°C]	%B3	%B4	B3:B4
1	120	332	731	2.50	1.7	537.5	66.3	33.6	1.97
2	114	278	605	3.17	1.7	532	66.1	33.9	1.95
3	145	372	755	2.85	3.1	518.8	64.7	35.3	1.83
4	128	319	658	2.73	3.5	503.6	69.2	30.7	2.25
5	104	259	593	2.73	3.8	517	69.5	30.5	2.28
6	111	268	565	2.84	3.2	516.6	64.3	35.7	1.80
7	113	299	638	2.48	12	494.1	66.4	33.5	1.98
8	122	329	676	2.63	3.7	545.6	69.3	30.7	2.25
9	119	302	626	2.83	1.5	471.6	63.9	36.1	1.77
10	154	380	742	2.43	16.9	487.2	76.4	23.6	3.24
11	130	337	677	2.94	2.7	551	64.3	35.7	1.80
12	111	291	647	2.47	2.2	532.6	65.9	34.1	1.93
13	118	319	697	2.90	1.9	482.1	63.8	36.2	1.76
14	151	407	769	2.57	1.8	499.5	66.7	33.3	2.00
15	135	367	725	2.64	2.7	516.6	57.6	42.4	1.36
16	104	261	574	2.77	3.2	504.7	61.0	39.0	1.57
17	134	334	666	2.88	1.9	527.6	61.4	38.6	1.59
18	147	373	734	2.68	2.8	563.6	63.4	36.6	1.73
19	124	302	589	2.80	15	490.9	67.2	32.8	2.05
20	120	334	656	2.63	2.4	541.3	70.4	29.5	2.38
21	98.4	218	458	2.81	2.1	535.3	61.4	38.6	1.59
22	132	330	658	2.66	1.8	563.8	64.0	36.0	1.78
23	136	377	728	2.74	2.3	495.1	64.0	36.0	1.78

3.3. Density

Density values ranged from 2.43 to 3.17 g/cm^2 (Figure 1A, Table 4). A quadratic model with statistical significance, as represented in Table 5, was developed to understand the relative influence of glass constituents, both individually and interactively, on glass density. Examination of the coefficients relating to the model suggests that factors contributing to an increase in the density response, ranked in order of magnitude, are B_2O_3*Na_2CO_3 > TiO_2 > ZnO > CaF_2*Na_2CO_3 > CaF_2 > ZnO*Na_2CO_3 > B_2O_3 > TiO_2*Na_2CO_3 > B_2O_3*CaF_2. Factors that contribute to a decrease in density, ranked in order of magnitude, are TiO_2*ZnO > B_2O_3*TiO_2 > Na_2CO_3 > CaF_2*TiO_2 > CaF_2*ZnO > B_2O_3*ZnO. A 3D surface plot (Figure 1A) representing the density response with respect to the variable concentrations of each glass constituent is also shown. The strongest correlation observed from the density data was an increase in density with decreasing B_2O_3 content (Figure 1B). Full model details, including coded coefficients and ANOVA outputs (R^2, adjusted R^2, predicted R^2, C.V. (%), and adequate precision), are provided in Table 5. The predicted R^2 is within reasonable agreement with the adjusted R^2 (i.e., the difference between the two values is less than 0.2), and the adequate precision (i.e., signal-to-noise ratio) is greater than 4 for every response, indicating the predictive power of the models.

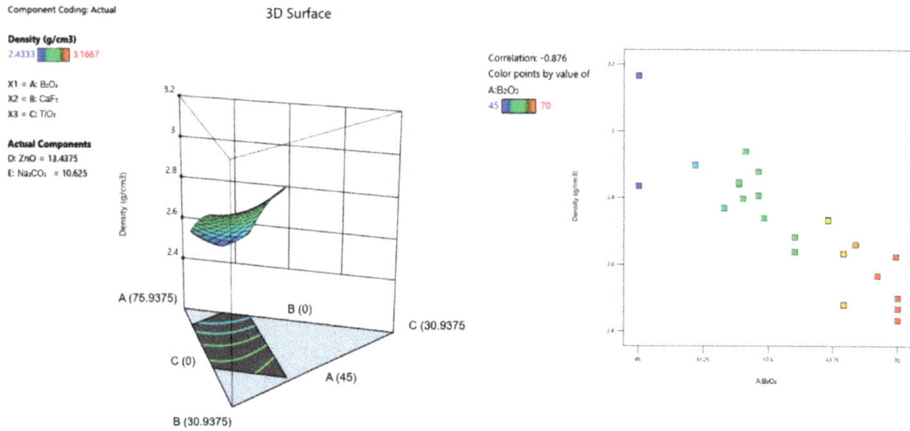

Figure 1. (**A**) 3D surface plot representing the density response across the compositional ranges of each constituent in the series. (**B**) Correlation graph of density vs. B_2O_3 content.

Table 5. Regression model coded equations based on L_Pseudo components for each response generated through Design-Expert. The summarized ANOVA including the R^2, adjusted R^2, predicted R^2, C.V. (%), and adequate precision is also included. The predicted R^2 for each response is within reasonable agreement to the adjusted R^2 (i.e., difference of <0.2) and the adequate precision values are greater than 4.

Response	Regression Model	R^2	Adjusted R^2	Predicted R^2	C.V. (%)	Adequate Precision
Density	$1.87 * B_2O_3 + 2.88 * CaF_2 + 5.05 * TiO_2 + 3.91 * ZnO + -0.66 * Na_2CO_3 + 0.73 * B_2O_3CaF_2 + -3.33 * B_2O_3TiO_2 + -0.21 * B_2O_3ZnO + 5.68 * B_2O_3Na_2CO_3 + -2.01 * CaF_2TiO_2 + -0.53 * CaF_2ZnO + 3.66 * CaF_2Na_2CO_3 + -3.46 * TiO_2ZnO + 1.07 * TiO_2Na_2CO_3 + 2.70 * ZnONa_2CO_3$	0.99	**0.98**	**0.83**	0.92	36.69
% Crystallinity	$10.08 * B_2O_3 + 34.51 * CaF_2 + 115.20 * TiO_2 + 24.56 * ZnO + 114.15 * Na_2CO_3 + -64.99 * B_2O_3CaF_2 + -50.99 * B_2O_3TiO_2 + -51.33 * B_2O_3ZnO + -157.37 * B_2O_3Na_2CO_3 + -284.31 * CaF_2TiO_2 + -57.53 * CaF_2ZnO + -203.37 * CaF_2Na_2CO_3 + -145.93 * TiO_2ZnO + -159.41 * TiO_2Na_2CO_3 + -186.40 * ZnONa_2CO_3$	0.99	**0.97**	**0.85**	17.08	27.01
T_g	$442.42 * B_2O_3 + 461.74 * CaF_2 + 569.54 * TiO_2 + 525.10 * ZnO + 673.09 * Na_2CO_3 + 489.77 * B_2O_3CaF_2 + -115.04 * B_2O_3TiO_2 + 194.12 * B_2O_3ZnO + -161.92 * B_2O_3Na_2CO_3 + 200.86 * CaF_2TiO_2 + 10.17 * CaF_2ZnO + -467.98 * CaF_2Na_2CO_3 + 65.39 * TiO_2ZnO + -535.48 * TiO_2Na_2CO_3 + -509.61 * ZnONa_2CO_3$	0.99	**0.97**	**0.91**	0.92	24.14
%B3	$94.53 * B_2O_3 + 60.04 * CaF_2 + 74.86 * TiO_2 + 75.23 * ZnO + 52.91 * Na_2CO_3 + -65.06 * B_2O_3CaF_2 + -12.24 * B_2O_3TiO_2 + -46.87 * B_2O_3ZnO + -54.87 * B_2O_3Na_2CO_3 + -9.13 * CaF_2TiO_2 + -0.91 * CaF_2ZnO + 13.00 * CaF_2Na_2CO_3 + -20.20 * TiO_2ZnO + -0.19 * TiO_2Na_2CO_3 + 15.02 * ZnONa_2CO_3$	0.99	**0.99**	**0.93**	0.59	60.79
%B4	$5.46 * B_2O_3 + 39.96 * CaF_2 + 25.14 * TiO_2 + 24.77 * ZnO + 47.08 * Na_2CO_3 + 65.06 * B_2O_3CaF_2 + 12.23 * B_2O_3TiO_2 + 46.87 * B_2O_3ZnO + 54.87 * B_2O_3Na_2CO_3 + 9.13 * CaF_2TiO_2 + 0.91 * CaF_2ZnO + -13.00 * CaF_2Na_2CO_3 + 20.20 * TiO_2ZnO + 0.19 * TiO_2Na_2CO_3 + -15.02 * ZnONa_2CO_3$	0.99	**0.99**	**0.93**	1.13	60.79
B3:B4	$4.65 * B_2O_3 + 2.24 * CaF_2 + 6.26 * TiO_2 + 2.44 * ZnO + 0.43 * Na_2CO_3 + -7.67 * B_2O_3CaF_2 + -3.86 * B_2O_3TiO_2 + -3.51 * B_2O_3ZnO + -4.85 * B_2O_3Na_2CO_3 + -7.82 * CaF_2TiO_2 + -0.82 * CaF_2ZnO + 2.17 * CaF_2Na_2CO_3 + -5.17 * TiO_2ZnO + -6.06 * TiO_2Na_2CO_3 + 1.31 * ZnONa_2CO_3$	0.99	**0.99**	**0.89**	2.17	54.56

3.4. X-ray Diffraction (XRD)

XRD analysis revealed varying degrees of crystallinity, ranging from 1.5% to 16.9% (Table 4). All glasses were free of identifiable crystalline species except samples #7, 10, and 19. The XRD spectra of these three samples showed sharp, visible peaks (Figure 2). Peak matching revealed that these crystalline peaks were characteristic of various titanium crystals (Table 6). Crystallinity data were used to derive a quadratic model with statistical significance (Figure 3, Table 5). The factors that contribute to an increase in crystallinity, ranked in order of magnitude, are TiO_2 > Na_2CO_3 > CaF_2 > ZnO > B_2O_3. Factors that contribute to a decrease in crystallinity, ranked in order of magnitude, are CaF_2*TiO_2 > CaF_2*Na_2CO_3 > ZnO*Na_2CO_3 > TiO_2*Na_2CO_3 > B_2O_3*Na_2CO_3 > TiO_2*ZnO > B_2O_3*CaF_2 > CaF_2*ZnO > B_2O_3*ZnO > B_2O_3*TiO_2. An additional finding with respect to the crystallinity of the glasses was an apparent effect between the TiO_2:CaF_2 concentration and crystallinity. To examine this effect, components B: CaF_2 and C: TiO_2 were removed from the axes of the 3D surface plot for crystallinity to permit visualization of the response surface upon adjusting these two constituents (Figure 4A–C). As shown in Figure 4A–C, the crystallinity of the glasses is at its maximum when the CaF_2 content is minimized (0 mol%) and the TiO_2 content is maximized (10 mol%). In contrast, the crystallinity is low when the CaF_2 content is maximized (20 mol%) and the TiO_2 content is maximized, or, similarly, if the CaF_2 content is maximized and the TiO_2 content is minimized (0 mol%). In addition, the difference between TiO_2 content and CaF_2 content was modelled to investigate whether there was a correlation between crystallinity and TiO_2–CaF_2 (Figure 4D).

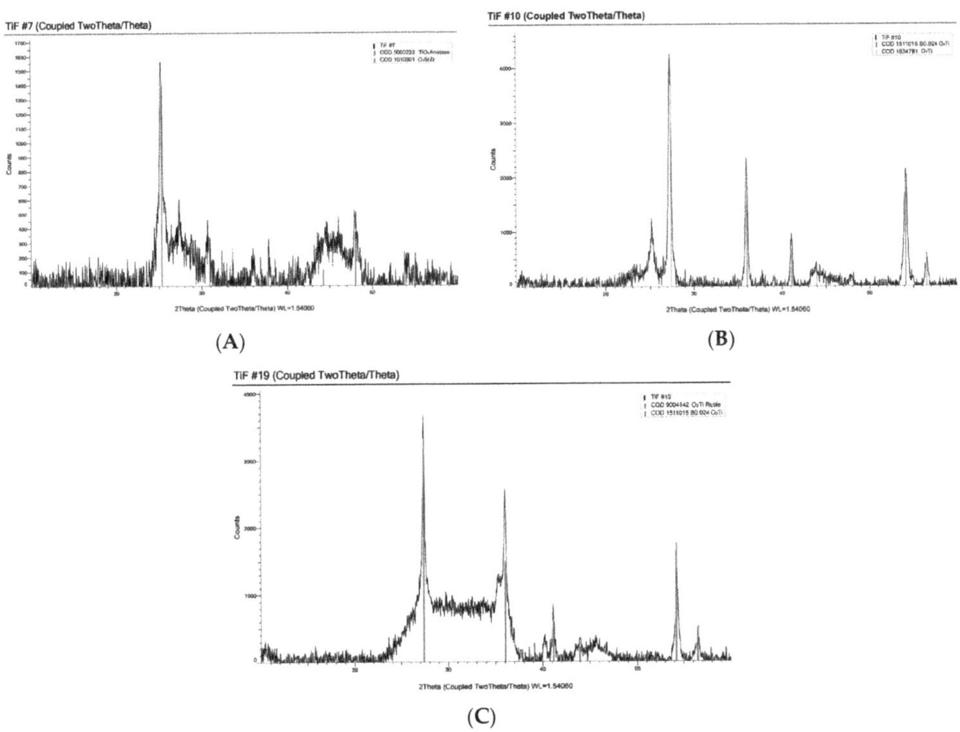

Figure 2. XRD crystal peak matching of samples (**A**) #7, (**B**) #10, and (**C**) #19.

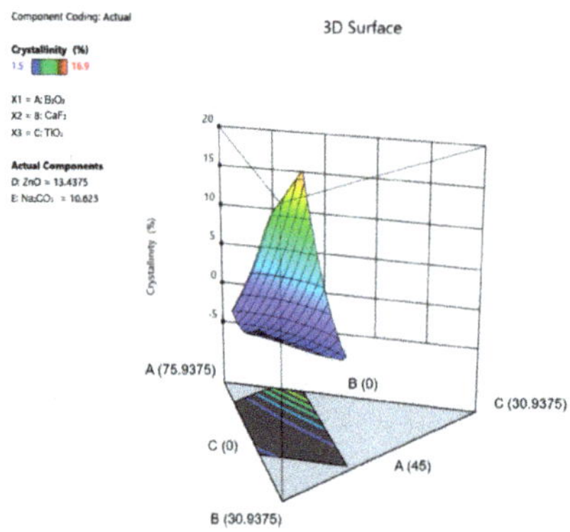

Figure 3. 3D surface plot representing the % crystallinity response across the compositional ranges of each constituent in the series.

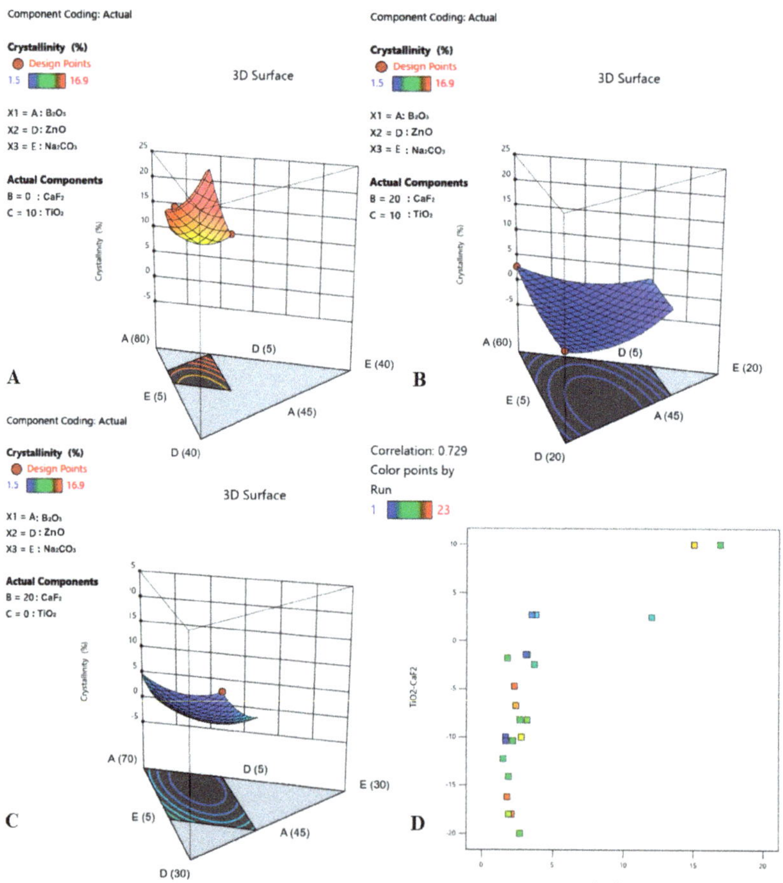

Figure 4. 3D surface plots representing the % crystallinity response for three distinct scenarios in which the concentrations of CaF_2 and TiO_2 vary but the other components remain constant ((**A**) Minimum CaF_2 and maximum TiO_2; (**B**) maximum CaF_2 and maximum TiO_2; (**C**) maximum CaF_2 and minimum TiO_2). (**D**) Correlation graph of crystallinity vs. TiO_2-CaF_2.

Table 6. Identification of crystals precipitated in samples #7, 10, 19.

Sample No.	COD ID	Chemical Name
7	COD 5000223	Titanium oxide—anatase
10	COD 1511015	Ti B0.024 O_2
	COD 1534781	TiO_2
19	COD 9004142	Titanium oxide—rutile
	COD 1511015	Ti B0.024 O_2

3.5. 11B MAS NMR

A representative ^{11}B MAS NMR line spectra is provided in Figure 5C. The peaks represent the B3 (12–20 ppm) and B4 ((−3)–5 ppm) units in the sample. The ^{11}B MAS NMR data for %B3 were fitted to a quadratic model with statistical significance (Table 5). The %B3 values ranged from 57.6 to 76.4% (Table 4). The strongest correlation observed from the ^{11}B MAS NMR data was an increase in the B4 fractions and a decrease in the B3 fractions with increasing quantities of CaF_2 (Figure 5A). Analysis of the coefficients derived from the coded equation for this response indicate that the glass constituents that contribute to an increase in %B3, ranked in order of magnitude, are B_2O_3 > ZnO > TiO_2 > CaF_2 > Na_2CO_3 > ZnO*Na_2CO_3 > CaF_2*Na_2CO_3. The glass constituents that contributed to a decrease in %B3, ranked in order of magnitude, are B_2O_3*CaF_2 > B_2O_3*Na_2CO_3 > B_2O_3*ZnO > TiO_2*ZnO > B_2O_3*TiO_2 > CaF_2*TiO_2 > CaF_2*ZnO > TiO_2*Na_2CO_3.

Similarly, the data for %B4 were fitted to a quadratic model with statistical significance (Table 5). The %B4 values ranged from 23.6 to 42.4% (Table 4). The factors that contribute to an increase in %B4, ranked in order of magnitude, are B_2O_3*CaF_2 > B_2O_3*Na_2CO_3 > Na_2CO_3 > B_2O_3*ZnO > CaF_2 > TiO_2 > ZnO > TiO_2*ZnO > B_2O_3*TiO_2 > CaF_2*TiO_2 > B_2O_3 > CaF_2*ZnO > TiO_2*Na_2CO_3. The factors that contribute to a decrease in %B4, ranked in order of magnitude, are ZnO*Na_2CO_3 > CaF_2*Na_2CO_3.

Lastly, analysis of the B3:B4 ratio (Figure 5B) produced a statistically significant quadratic model (Table 5), with values ranging from 1.358 to 3.24 (Table 4). The factors that contributed to an increase in B3:B4, ranked in order of magnitude, are TiO_2 > B_2O_3 > ZnO > CaF_2 > CaF_2*Na_2CO_3 > ZnO*Na_2CO_3 > Na_2CO_3. The factors that contributed to a decrease in B3:B4, ranked in order of magnitude, are CaF_2*TiO_2 > B_2O_3*CaF_2 > TiO_2*Na_2CO_3 > TiO_2*ZnO > B_2O_3*Na_2CO_3 > B_2O_3*TiO_2 > B_2O_3*ZnO > CaF_2*ZnO.

Additionally, the experimental fraction of trigonally coordinated boron was plotted against the intended B_2O_3 concentration in the glasses (Figure 5D). The 2% error bars (vertical axis) account for potential variations in the NMR data when determining the relative fraction of B3. As shown in Figure 5D, the data points are very scattered, and therefore, the correlation between B_2O_3 content and boron coordination is somewhat ambiguous. To investigate this scattering further, NMR spectra for samples with identical B_2O_3 concentrations but varying concentrations of additive components were overlayed to compare the B3 and B4 sites (e.g., position and shape) and investigate the impact of varying modifiers on glasses with constant B_2O_3 content. Two representative overlay plots are shown in Figure 6A,B, wherein the B3:B4 ratios clearly varied across samples. Although some variation between samples was expected due to differences in the modifying elements and their concentrations, these deviations warranted further investigation.

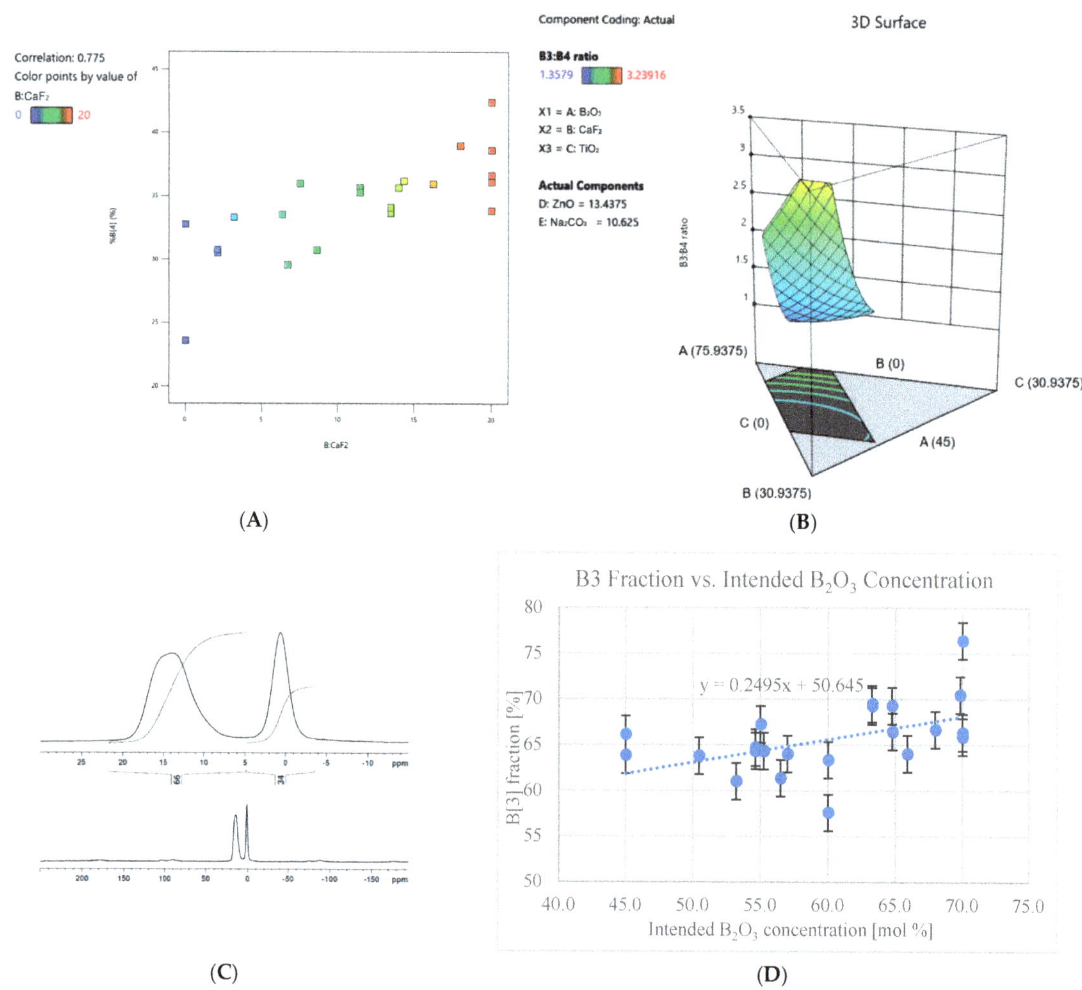

Figure 5. (**A**) 3D surface plot depicting the ratio of B3:B4 structural units relative to varying amounts of each constituent in the series. (**B**) Correlations graphs of %B4 vs. CaF_2 content. (**C**) Representative ^{11}B MAS NMR spectra for the glasses under investigation (sample #1). (**D**) Experimentally obtained concentration of B3 versus intended B_2O_3 concentration for each glass in the series.

Because these glasses are highly complex, consisting of several different cations as well as oxygen and a fluoride, a typical B3 to O:B correlation could not be used. Alternatively, the B3 content was plotted against the overall cation charge per boron atom based on the intended and actual compositions (Figure 7). For pure B_2O_3, the charge per boron is 3, and one would expect to see 100%B3. Similarly, for tetrahedrally coordinated boron, one would see 0%B3. Based on this, a slope of −100%/charge is expected. However, as shown in Figure 7, the slopes correlating with the raw NMR data were −4.6 and −4.3, respectively.

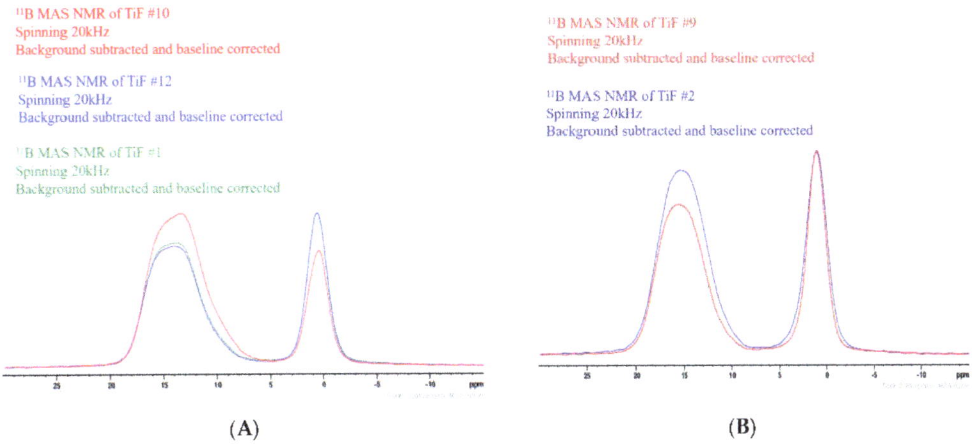

Figure 6. Overlay plots of glass samples having the same concentration of B_2O_3, but varying concentrations of additive elements. (**A**) Overlay plot of sample #1, 12, 10, all having 70 mol% B_2O_3 (note: samples #1 and 12 are replicate compositions). (**B**) Overlay plot of sample #9 and 2, both having 45 mol% B_2O_3.

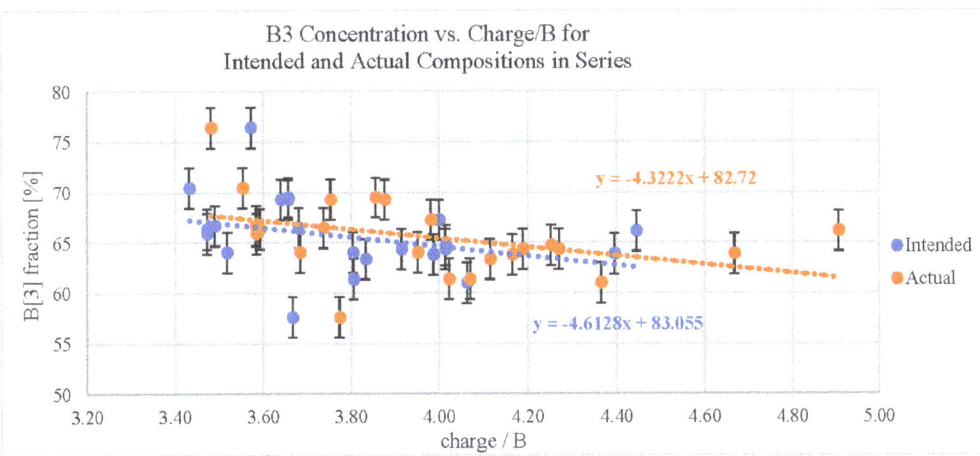

Figure 7. Fraction of B3 versus the overall cation charge per boron atom based on the intended composition of each glass (blue) and the actual, experimental composition of each glass (orange).

3.6. Differential Scanning Calorimetry

T_g values ranged from 471.6 °C to 563.8 °C (Table 4). These values were fitted to a quadratic model with statistical significance (Figure 8A, Table 5). Analysis of the coefficients derived from the coded equation describing this model indicates that the glass constituents contributing to an increase in T_g, ranked in order of magnitude, are Na_2CO_3 > TiO_2 > ZnO > B_2O_3*CaF_2 > CaF_2 > B_2O_3 > CaF_2*TiO_2 > B_2O_3*ZnO > TiO_2*ZnO > CaF_2*ZnO. Constituents that contribute to a decrease in T_g, ranked in order of magnitude, are TiO_2*Na_2CO_3 > ZnO*Na_2CO_3 > CaF_2*Na_2CO_3 > B_2O_3*Na_2CO_3 > B_2O_3*TiO_2. The strongest correlation observed from DSC data was an increased T_g with decreasing Na_2CO_3 content (Figure 8B).

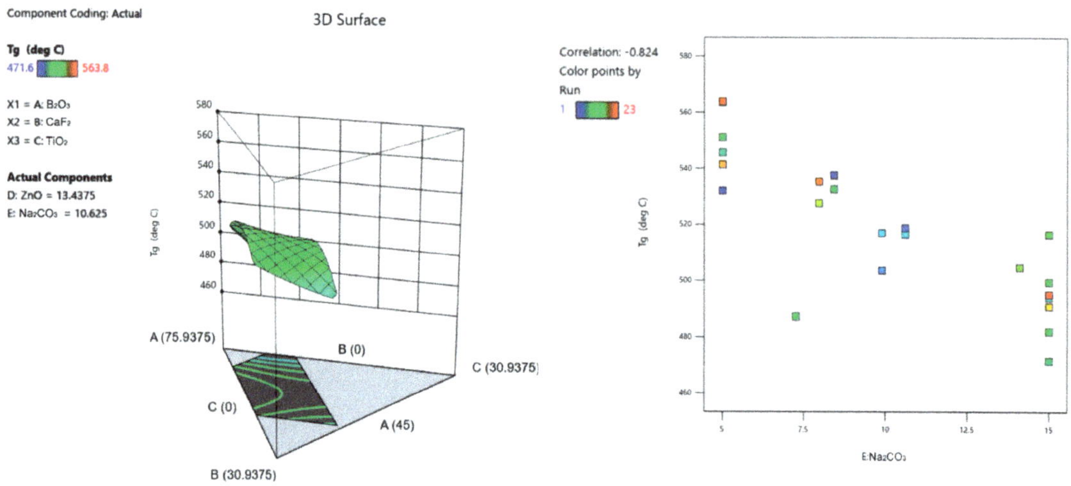

Figure 8. (**A**) 3D surface plot representing the T_g response across the compositional ranges of each constituent in the series. (**B**) Correlation graph of T_g vs. Na_2CO_3 content.

As shown by the coefficients in the coded equations in Table 5, the individual and interaction terms associated with a given constituent can have opposing impacts. To provide enhanced clarity on the overall impact of a single constituent on a given response, a graphical correlation table is depicted in Figure 9.

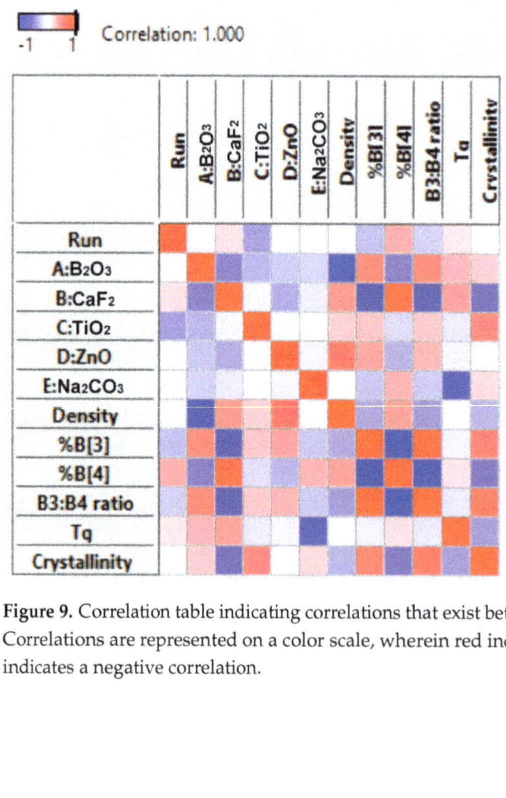

Figure 9. Correlation table indicating correlations that exist between glass constituents and responses. Correlations are represented on a color scale, wherein red indicates a positive correlation and blue indicates a negative correlation.

3.7. Optimization and Validation

The two optimal glass formulations derived from the optimization criteria outlined in Table 3 are shown in Table 7. The predicted values for density, %B3, %B4, T_g, as well as desirability scores for each optimized formulation, are also included in Table 7.

Table 7. Composition of optimized formulations and predicted values for select parameters.

Glass	B_2O_3	CaF_2	TiO_2	ZnO	Na_2CO_3	Density [g/cm^3]	%B3	%B4	T_g [°C]	Desirability
A	70.0	8.1	3.2	13.7	5.0	2.5	70.7	29.2	539.9	0.902
B	51.9	20.0	3.5	14.8	9.81	2.9	62.2	37.8	510.0	0.942

XRD analysis confirmed that both optimized glasses were free from identifiable crystalline species. The actual experimental values derived from density, DSC, and ^{11}B MAS NMR analyses of optimized glasses A and B are listed in Tables 8 and 9, respectively, as well as the predicted values of the model and the 95% prediction interval lower and upper limits. As shown in Tables 8 and 9, the actual values are within the 95% prediction interval for density and T_g. However, the results for %B3 and %B4 fall outside the model's prediction interval, indicating that the %B3 and %B4 models are not predicting as accurately as desired.

Table 8. Predicted and actual values for density, T_g, %B3, and %B4 responses for optimized glass A. Actual values that fall within the 95% prediction interval indicate that the model for that given response is predicting accurately at the optimized settings. Actual values that do not fall within the 95% prediction interval are highlighted in red.

Optimized Glass A				
Response	Predicted	95% Prediction Interval Low	Actual	95% Prediction Interval High
Density [g/cm^3]	2.5	2.428	2.548	2.571
T_g [°C]	539.9	526.464	534.2	553.602
B3 [%]	70.75	69.637	69	71.853
B4 [%]	29.25	28.147	31	30.363

Table 9. Predicted and actual values for density, T_g, %B4, and %B3 responses for optimized glass B. Actual values that fall within the 95% prediction interval indicate that the model for that given response is predicting accurately at the optimized settings. Actual values that do not fall within the 95% prediction interval are highlighted in red.

Optimized Glass B				
Response	Predicted	95% Prediction Interval Low	Actual	95% Prediction Interval High
Density [g/cm^3]	2.9	2.831	2.874	2.971
T_g [°C]	510	496.5	511	523.159
B3 [%]	62.19	61.118	60	63.296
B4 [%]	37.81	36.704	40	38.882

4. Discussion

The non-linear behavior exhibited in borate glass systems is referred to as the borate anomaly and is the reason why the structure and properties of borate glasses with modifier fractions greater than 30 mol% are very difficult to predict [11,16]. The difficulty in predicting the composition–structure–property relationships of borate glasses in this high modifier fraction range introduces a significant hurdle in advancing the optimization of glass compositions beyond conventional limits. In essence, addressing the challenges posed by the borate anomaly has the potential to unlock new avenues for innovation in glass science and broaden the scope of materials with enhanced functionalities.

Recognizing the limitations of traditional trial-and-error or OVAT approaches in understanding the constituent interactions that occur within such complex glass systems has encouraged a shift towards adopting more advanced methodology. As such, a DoM approach was taken for this investigation. This approach stands out for its ability to discern both the individual and interaction effects of each glass constituent on material properties, thus offering a more profound understanding of the convoluted relationships that exist within a glass system. Moreover, it permits rapid screening across broad compositional ranges, thus directly supporting the accelerated design, development, and deployment of novel glass materials. The glass system under investigation in this work was designed (from a compositional standpoint) within an interesting region of the borate anomaly where structural behavior is not well defined [11]. Specifically, these glasses have modifier fractions ranging between 30 and 55 mol%. While this region offers the opportunity to experiment within a wide range of structural constraints, the variability in network properties makes it extremely difficult to predict. This challenge is exacerbated by the inclusion of multiple elements belonging to various groups (i.e., groups 1, 2, 4, 12, 13, 17) and exhibiting variable physical properties.

As shown in Table 5, the DoM statistical modelling approach has permitted the generation of polynomial equations with coded coefficients that indicate the relative influence of each constituent on variable responses. These equations are dependent on the design constraints listed in Table 1, wherein the concentrations of each constituent (in mol%) are constrained to add to a fixed total (i.e., 100 mol%), meaning that the value of a given constituent is necessarily dependent on the values of the others. These polynomial equations are extremely valuable predictive tools and clearly illustrate that the complex interactions that occur between glass constituents can have massive implications on material structure and properties. To gain insight into whether a given constituent is causing an increase or decrease in a response, the polynomial equation should be used in conjunction with the correlations table (Figure 9) and the 3D surface plots. The correlations table is a direct visualization of the inputs versus the outputs (i.e., data versus data), independent of any model. This table is extremely valuable in identifying any correlations that exist between two properties or between property and composition. Alternatively, the 3D surface plots are a visual representation of a given response relative to the variable concentrations of each constituent in the glass. The 3D surface plots alone demonstrate the non-linearity and unpredictability of relationships within these complex glass systems. This graphical representation not only showcases the complexity of the system but also highlights the utility of predictive modelling approaches. Specifically, these systematic methods simultaneously predict the individual and interaction effects of multiple constituents on material properties over vast compositional ranges, providing a comprehensive understanding that transcends linear interpretations.

As shown in Figure 1B, there was a linear correlation observed between density and B_2O_3 content in these glasses, wherein the density increased with decreasing B_2O_3 content. This behavior can also be seen in the 3D surface plot representing the density response (Figure 1A), wherein the surface contour is at its lowest point and component A: B_2O_3 is at its maximum. This trend may be explained by the increasing fraction of modifying elements that is associated with a decreasing B_2O_3 content, as these modifying elements have higher atomic masses compared to boron (e.g., the atomic mass of B is 10.811 g/mol while the atomic masses of Ti, Zn, Ca, and F are 47.867, 65.38, 40.078, and 18.998 g/mol, respectively). Similar findings have been reported in the literature with respect to the effect of ZnO [34] and TiO_2 [35] on increasing the density of borate glasses. This is further supported by the relative impact of each modifying element on density, as per the coded equation in Table 5, wherein the constituents with larger atomic masses (e.g., TiO_2 and ZnO) had the greatest individual impacts on increasing density, as would be expected. Furthermore, titanium and zinc are classified as intermediate elements according to Sun and Dietzel's classifications, meaning they can act as glass network modifiers or glass network formers depending on the glass composition [16]. For instance, titanium has been shown to exist in $[TiO_4]$, $[TiO_5]^{2-}$, and $[TiO_6]^{2-}$ coordination states, wherein it assumes different roles based on its structure and the type of glass network [36]. Specifically in borate glass networks, titanium has been reported to exist as a network former in the TiO_4 tetrahedral state and a network modifier in the $TiO_6{}^{2-}$ octahedral state [26,29,30,37–39]. Similarly, in borate glass networks, zinc has been shown to act as a network former in the fourfold coordination state and a network modifier in the sixfold coordination state [40–45]. Therefore, it is possible that these constituents increased the density of these glasses by (1) assuming a network-forming role, thus contributing to building the basic framework of the glass, or (2) acting as a network modifier and forming crosslinks between existing borate units, thus increasing the network connectivity of the glass. For instance, in a borate glass network, the addition of TiO_2 up to 15 mol% was reported to cause a substitution of $BO_3{}^{3-}$ units with Ti^{4+} units due to the small ionic radius and large electric charge of Ti^{4+} ions [46]. This resulted in the formation of B–O–Ti bonds rather than B–O–B bonds and was reported to cause increased network connectivity. Alternatively, a study by El-Falaky and Guirguis found that the addition of ZnO up to 35 mol% to a borate glass caused a decrease in tetrahedrally coordinated boron, as ZnO_4 assumed the role of a network former and gradually replaced borate units. This resulted in an increase in the number of bonds per unit volume, an increased packing density, and a decreased molar volume [44]. These findings reiterate how glass structural behavior varies depending on the number of modifying elements, the properties of these modifying elements, and the concentrations of each element in the network, further emphasizing the importance of understanding the influence that each constituent has on material properties, both individually and interactively.

Another interesting finding observed in this work was the correlation between the appearance of quenched glass samples and their measured % crystallinities. The quenched glass samples varied in appearance, some being completely transparent, and others with varying fractions of white, opaque coloring mixed throughout. XRD analysis confirmed the absence of identifiable crystalline species in all samples, except for samples # 7, 10, and 19. These three samples had the largest fractions of white opaque coloring and had the highest % crystallinities, according to XRD analysis. Peak matching of the XRD spectra pertaining to these three samples revealed the presence of various titanium crystals, such as anatase and rutile, indicating that titanium is likely phase separating into crystalline phases. Similar findings have been reported in the literature with respect to the role of titanium in inducing crystallization in various glasses and glass ceramics [36,38,46–51]. In fact, titanium is widely acknowledged as an effective nucleating agent and is believed to act as a catalyst for initiating crystallization centers that induce controlled bulk crystallization and subsequent volume crystallization [38]. Interestingly, as shown in Figure 4, there appeared to be an effect between the CaF_2/TiO_2 ratio and the crystallinity of the glasses, wherein samples with the greatest % crystallinities included higher concentrations of TiO_2 compared to CaF_2. This effect was further supported by the correlation between crystallinity and TiO_2–CaF_2, wherein an increased % crystallinity was correlated with an increased difference between TiO_2 and CaF_2. This suggests that the presence of CaF_2 may have hindered the formation of titanium crystalline phases within the glasses. The potential Ti crystalline phase separation in the glasses under investigation was further evidenced through ^{11}B MAS NMR findings.

The ^{11}B MAS NMR data in Figure 5D show the percentage of trigonally coordinated boron plotted against the intended B_2O_3 content in the glasses. Contrary to what is known in the binary borate glass literature with respect to changes in boron coordination with varying the B_2O_3 content, there were no identifiable trends between the B3 fraction and the B_2O_3 content for these glasses. Furthermore, the data points in Figure 5D appeared to be very scattered. To investigate the cause of the scattering, the NMR spectra for samples with identical B_2O_3 concentrations but varying concentrations of additive components were overlayed to compare the position and shape of the B3 and B4 sites (representative plots shown in Figure 6A,B). These overlay plots, in combination with Figure 5B, demonstrate how, even across samples having identical B_2O_3 content, the B3:B4 ratios varied, which can be attributed to the differences in modifier type and concentrations. For instance, these variations may be a result of some of the Ti^{4+} being exchanged with cations with +2 or +1 charge. Such substitutions could disrupt the charge balance in the network, leading to changes in the overall structure, connectivity, and stability of the glass. For example, the substitution of a Ti^{4+} cation with a Na^+ cation may result in a decrease in O–Ti–O crosslinks between borate structural groups and an increased formation of non-bridging oxygens, wherein the smaller Na^+ network-modifying cations fit more easily into the cages/sites of the existing network, and the Ti^{4+} cations may preferentially assume a network-forming role in forming crosslinks between borate units. A graphical representation illustrating the potential structural modifications induced by the inclusion of network modifiers, such as Na^+, can be found in previous work [11]. Furthermore, compositional differences between intended and actual concentrations arising from changes during the melting process could potentially contribute to the observed deviations. However, the change in cation charges and the small compositional changes that could arise between intended and actual compositions under normal circumstances are not sufficient to fully explain the scatter in the data. Similar deviations were observed when the B3 content was plotted against the overall cation charge per boron atom based on intended and actual compositions (Figure 7). Evidently, both plots show trendlines far from the predicted slope of −100%/charge and are almost flat. Two possible occurrences that may explain the deviations from expected behavior are as follows: (i) The sample composition changed dramatically from the intended composition, and (ii) phase separation is occurring within the glasses. With regards to the former, one possible cause for differences in the intended versus actual compositions is that the majority of fluorine included in these glasses was lost during glass synthesis. It is possible that BF_3 formed from the initial components, which is a gaseous compound at room temperature and has a boiling point of −100.3 °C. If BF_3 burned off during melting, this would result in a decreased concentration of both B and F (and potentially other components) in the samples, thus altering the content of B3 and B4 units in the network. With regards to the latter, phase separation could explain the significant variations observed in the shape and position of the boron signal for 3 and 4-fold coordinated sites in these samples. Specifically, if a non-boron-containing crystal formed, the relative concentration of boron in the remaining glassy matrix would be higher than anticipated. Conversely, if it were a boron-containing crystal, the relative concentration of boron in the glassy phase would be lower than predicted. These differences in boron content throughout the crystal and glassy phases and, consequently, the variations in boron content between predicted and actual concentrations could lead to the fluctuations in %B3 and B4 signals observed in these glasses.

An additional factor that is likely to contribute to the confounding nature of the borate network structure in these glasses is the inclusion of two intermediate elements, TiO_2 and ZnO. Interestingly, the coefficients for ZnO and TiO_2 with respect to %B3 (and %B4) were very similar (e.g., +75.23 * ZnO and +74.86 * TiO_2), indicating that these constituents had a similar impact on this response, despite their different cationic charges and field strengths. Based on Sun's single bond strength criterion, Ti and Zn have very similar bond strengths (i.e., 73 and 72 kcals, respectively), indicating that both elements may assume similar roles within the network. Furthermore, these two elements have the highest cationic field strengths of all additive elements in the glasses, indicating that they may dictate structural changes to satisfy their own requirements. For example, the inclusion of TiO_2 may favor the substitution of B–O–B bonds for B–O–Ti bonds, wherein Ti^{4+} forms ionic crosslinks between borate units. Alternatively, titanium may have disrupted the borate network by forming $[TiO_6]^{2-}$ octahedrons in which non-bridging oxygens were increasingly formed, and as a result, the fraction of tetrahedrally coordinated boron decreased. The effect of titanium on boron coordination in a titanium- and fluorine-containing borate glass was previously investigated, wherein the increase in TiO_2 content up to approximately 25 mol% was associated with a decrease in the proportion of tetrahedrally coordinated boron, thus indicating a transformation from B4 to B3 structural units in regions enriched with TiO_2 [52]. The following Ti coordination change was thought to be the catalyst for this transformation: $TiO_{4/2} + 2B(O,F)_4 \rightarrow Ti(O,F)_6 + 2BO_3$. FTIR analysis findings suggested that this coordination change was associated with an increased presence of network-forming TiO_4 units at low concentrations of TiO_2 (up to ~10 mol% TiO_2) and network modifying TiO_6^{2-} units at higher concentrations of TiO_2 (~13 to 20 mol% TiO_2). In addition, similar findings were reported by Lakshmi and Cole [53] in a separate titanium and fluorine-containing borate glass. Specifically, FTIR analysis revealed that with increasing TiO_2 content (up to 0.7 mol%), the intensity of the band assigned to TiO_6 octahedral units increased, while the reverse trend ensued for the band assigned to TiO_4 tetrahedral units. Similarly, with increasing TiO_2 content, the intensity of the band assigned to BO_3 units grew at the expense of the BO_4 unit band. Based on these findings, the authors suggested that titanium assumed the role of a network modifier in the glasses.

With respect to the role Zn has in modulating borate glass structure, it too can enter the glass network in various coordination states, thus altering the network differently depending on its structure. When functioning as a network modifier, Zn^{2+} ions disrupt B–O–B bonds, thus decreasing the number of bridging oxygens in the network. Conversely, when zinc assumes a network former role, it can take the form of ZnO_4 structural units, wherein zinc is covalently bonded with four oxygen ions [43]. In a borate glass, the addition of ZnO up to 35 mol% was shown to decrease the boron coordination number as ZnO_4 units replaced boron atoms in the network [44]. Similarly, for binary zinc–borate glasses, Topper et al. [34] reported that with the addition of ZnO, there was an increasing population of high field strength tetrahedral Zn^{2+} units, which led to the depolymerization of larger borate arrangements due to the increasing number of high charge density borate anions. The authors noted, however, that the increased number of strong Zn^4–O–B crosslinks partially offset the disruption of the B–O–B bonds. Furthermore, in a separate study on zinc-containing borate glasses, Cetinkaya et al. found that in samples containing greater than 5 wt. % ZnO, zinc acted as a network former. This finding was based on the presence of bands belonging to ZnO_4 structural units in samples containing greater than 5% ZnO and the absence of these bands otherwise. In addition, the relative intensities of peaks corresponding to bridging oxygens were lower for sample BZ5 (5 wt.% ZnO) and increased for samples BZ10 and BZ15 (10 and 15 wt.% ZnO), thus further supporting the above-mentioned findings.

Based on the coefficients of the coded equations generated from ^{11}B MAS NMR analysis, one could assume that, directionally, both TiO_2 and ZnO are primarily contributing to an increase in %B3. However, this relationship is not so straight forward, as there are no strong correlations that exist between either of these constituents and boron coordination, and the 3D surface plots show variable responses depending on the relative amounts of the other components. Despite gaining directional information on the influence of these TIIs, the impact of both intermediate elements on boron coordination is somewhat unclear, and further studies may be necessary to ascertain the complex structural relationships that exist in a multi-component borate glass network containing two intermediate elements.

The strongest correlation observed for the boron coordination was an increased concentration in %B4 (and a decrease in %B3) with increasing CaF_2 content (Figure 5A). This is evidenced by the 3D surface plot for the B3:B4 ratio (Figure 5B), wherein the ratio is larger where component B:CaF_2 is at a minimum. According to the coded equation for %B4 (Table 5), CaF_2 also had the largest individual impact on increasing the response. These findings collectively suggest that CaF2 plays a significant role in modulating the borate network structure. However, the way this constituent affects the network is complex, owing to the presence of both fluorine and calcium in the starting reagent. In the context of fluorine, it has been demonstrated, through NMR analysis of aluminosilicate glasses, that fluorine ions can replace bridging oxygens around silicate tetrahedral units, thereby creating non-bridging, dangling bonds and causing a significant reduction in the glass transition temperature [46]. Similarly, in the borate glass literature, fluorine ions have been shown to substitute the oxygen around trigonal boron atoms due to the similarity in ionic radii between oxygen and fluorine [37]. This substitution results in the formation of non-bridging oxygens and thus increases the fraction of B3 units, contrary to what was observed in this glass system. Alternatively, additional studies have demonstrated fluorine's preference for forming bonds with other modifier elements present in the network, wherein the replacement of bridging oxygens by nonbridging fluorine ions was not detected [54]. While the mechanism by which fluorine may have induced the increased formation of B4 units in these glasses is not entirely clear, it is important to consider the possible discrepancies between the intended and actual fluorine content in these glasses. As previously discussed, and evidenced through ^{19}F MAS NMR analysis, there was an absence of fluorine signal in the samples, indicating that the majority of fluorine may have been burned off in the melting process in gaseous BF_3 form. If this were in fact the case, it is possible that the correlation between %B3 and CaF_2 was primarily a result of the calcium. In this regard, Ca^{2+} ions may have increased the conversion of boron from a 3-coordinated state to a 4-coordinated state by fitting into the interstices of the borate network and balancing the negative charges on neighboring BO_4^- groups [16]. As a result of this interaction, the network's connectivity and viscosity increase. Furthermore, in binary borate glass networks, the addition of CaO up to a maximum of approximately 40 mol% has been shown to increase the glass network connectivity and increase the formation of BO_4^- tetrahedral units at the expense of BO_3 units, whereafter the reverse trend ensues [55,56].

In conjunction with density, XRD, and NMR analyses, the thermal characteristics of the glass networks were also assessed. When comparing the relative effects of each glass constituent on the glass transition temperature of the materials, the individual term with the largest coefficient, and thus the greatest impact on T_g, was Na_2CO_3 (Table 5). This coefficient is positive, indicating that Na_2CO_3 contributed to an increase in T_g. However, to understand the overall impact of Na_2CO_3 on T_g, the interaction terms must be considered. Interestingly, the coefficients corresponding to interaction terms containing Na_2CO_3 were all negative, indicating that Na_2CO_3 in combination with the other constituents had a relatively greater negative impact on the glass transition temperature (i.e., contributed to a decrease in T_g). This was further supported by the strong correlation observed between T_g and Na_2CO_3 content, wherein there was a decreased T_g value with increasing Na_2CO_3 content (Figure 8B). Furthermore, this effect was verified by adjusting the concentration of component E: Na_2CO_3 in the 3D surface plot, wherein the surface clearly shifted to higher temperatures at the lowest Na_2CO_3 content and vice versa. These combined findings demonstrate the complex role that sodium may have in these networks and how that role may differ when in combination with other constituents. This is a great example of why it is crucial to understand not only the individual effects of glass constituents, but also the complex, and evidently influential, synergistic effects that may occur between them. With regards to the mechanism by which Na_2CO_3 may cause a decrease in glass transition behavior, sodium has previously been used as an effective flux in glass materials, thus increasing the ease of melting and homogenization processes [57]. This is because, past a certain concentration, the introduction of electropositive cations of alkali metals, such as Na^+, is known to break up the interlinked borate network by converting bridging oxygen atoms into non-bridging oxygen ions [58]. This conversion decreases the network connectivity of the glass and thus alters its thermal characteristics, such as viscosity and glass transition temperature. However, based on the principles of the borate anomaly in binary Na_2O–B_2O_3 glasses, increasing the sodium content initially induces the formation of BO_4^- tetrahedral units at the expense of BO_3 units until the fraction of modifier hits a maximum, whereupon non-bridging oxygens begin to form [11]. Because (1) the glasses under investigation contain modifier fractions ranging from 30–55 mol% that include variable concentrations of alkaline, alkaline–earth, transition metal, and halogen elements, and (2) there were no evident correlations between Na_2CO_3 content and %B4 and %B3, it is difficult to ascertain the role of Na_2CO_3 in modulating B3 and B4 fractions and how that might affect the thermal properties of these glasses. Although the boron coordination is often assumed to be correlated to the glass transition temperature [59], there was no correlation between the glass transition temperature and %B3 or %B4 in these glasses. These results indicate that the mechanism by which Na_2CO_3 caused a decrease in T_g was not related to the relative fractions of B3 and B4 units in the network. As such, an alternative theory for the decreased T_g values is that, upon increasing the Na_2CO_3 content, B–O bonds are increasingly substituted by lower bond dissociation energy Na–O bonds in the network. For instance, the addition of Na_2O up to 50 mol% has previously been shown to decrease the glass transition temperature in Na_2O–ZnO–B_2O_3 glasses, wherein authors attributed this decrease to the low bond association energy of Na–O bonds (257 kJ/mol) compared to B–O bonds (806 kJ/mol) [60].

Among the many benefits of employing predictive modelling approaches in the design and development of novel glass materials, perhaps one of the most impactful is the ability to predict and optimize new formulations tailored to specific properties. Optimization studies offer a systematic and strategic approach to enhancing the functionality and performance of novel materials. These studies allow researchers to fine-tune compositions with precision, ensuring that the material properties align optimally with desired outcomes. By leveraging surface regression methodologies, we can not only navigate the complexities of multidimensional systems but also discover optimal solutions within vast compositional spaces using limited time and resources. By identifying the most favorable compositions and configurations, these studies streamline the materials discovery process, paving the way for quicker translation of novel materials into practical applications. In the current investigation, the optimization and validation study was used to assess the predictive capabilities of the models by comparing the measured properties of optimized solutions against the predicted values of the model. As shown in Tables 8 and 9, the actual values derived from density and DSC analyses of the two optimized glasses are within the 95% prediction interval, confirming that the density and T_g models are predicting accurately at both optimized settings. However, the results arising from ^{11}B MAS NMR analysis fell outside the model's 95% prediction interval, indicating that the %B3 and %B4 models were not predicting as accurately as desired. Although the models generated for %B3 and %B4 and B3:B4 were statistically significant, the ^{11}B MAS NMR results (specifically the observed data scattering and deviation from expected behavior) suggested that the changes in borate network structure were difficult to accurately predict in these systems. This was somewhat expected given the unpredictable nature of the borate glass structure within the anomalous region. In light of these findings, it is worth noting that the measured values for %B3 and %B4 were provided with a ± 2 to account for measurement error/variability in chemical shift. If the ± 2 error is applied to the actual values pertaining to %B3 and %B4, there is a scenario in which these values fall within the 95% prediction interval for both optimized glasses. An additional consideration is that the optimization study was limited by conducting only one confirmation run to confirm the accuracy of the models. In future work, both optimized formulations should be synthesized and characterized at least three times to account for any possible variations in glass synthesis, measurement error, or human error. Ultimately, the results of the current optimization study demonstrate that by employing a predictive modelling approach, it was possible to accurately predict the individual and interaction effects of various glass constituents on the density and thermal properties of these complex, multi-component glasses. Although the optimization and validation results suggested that the models generated from ^{11}B MAS NMR analysis were not predicting as accurately as desired, the models still provide directional information on the individual and interaction effects of each constituent on the borate network structure.

5. Conclusions

The application of a predictive modelling approach, based on Design of Mixtures (DoM), has successfully elucidated the individual and interaction effects of B_2O_3, CaF_2,

TiO_2, ZnO, and Na_2CO_3 on the structure and properties of oxyhalide borate glasses with modifier fractions $\geq$ 30 mol%. This study highlights the non-linear and complex relationships that exist within multi-component glass systems, particularly within the anomalous region of borate glasses. Subsequent to developing statistically significant models for various material characteristics, the authors illustrated how to predict and optimize new glass compositions tailored for specific properties, offering new insights into the design and development of soluble glass networks. Although some challenges in predicting changes in the borate network structure are acknowledged, the study demonstrates the effectiveness of systematic predictive modelling in overcoming traditional barriers in glass science. Such advancements will open new avenues for the accelerated discovery and development of novel glasses for a wide range of applications. Future work will focus on further elucidating the observed phenomena, particularly the scattering in ^{11}B MAS NMR data and the potential loss of fluorine during synthesis, to refine our predictive capabilities and expand our understanding of these complex materials.

Author Contributions: Conceptualization: B.K. and D.B.; methodology: B.K. and D.B.; investigation: B.K.; formal analysis: B.K. and D.B.; writing—original draft: B.K.; writing—reviewing and editing: B.K. and D.B.; visualization: B.K.; project administration: B.K. and D.B.; validation: B.K.; resources: B.K. and D.B., supervision: D.B.; funding acquisition: D.B. All authors have read and agreed to the published version of the manuscript.

Funding: This research was funded by Natural Sciences and Engineering Research Council of Canada (NSERC) discovery grant number 2022-03371.

Institutional Review Board Statement: Not applicable.

Informed Consent Statement: Not applicable.

Data Availability Statement: The original contributions presented in the study are included in the article, further inquiries can be directed to the corresponding author.

Acknowledgments: We acknowledge (i) the support of the Dalhousie Medical Research Foundation (DMRF), (ii) the technical support of Ulrike Werner-Zwanziger who generated the ^{11}B MAS NMR data for our analysis, and (iii) Alicia Oickle for ongoing technical support and assistance.

Conflicts of Interest: The authors declare no conflicts of interest.

References

1. Hoppe, A.; Boccaccini, A.R. Biological Impact of Bioactive Glasses and Their Dissolution Products. *Front. Oral Biol.* **2015**, *17*, 22–32. [CrossRef]
2. Place, E.S.; Evans, N.D.; Stevens, M.M. Complexity in biomaterials for tissue engineering. *Nat. Mater.* **2009**, *8*, 457–470. [CrossRef]
3. Mouriño, V.; Cattalini, J.P.; Boccaccini, A.R. Metallic ions as therapeutic agents in tissue engineering scaffolds: An overview of their biological applications and strategies for new developments. *J. R. Soc. Interface* **2012**, *9*, 401–419. [CrossRef]
4. MacDonald, K. Strontium Releasing Composite Resin Cements. Ph.D. Thesis, Dalhousie University, Halifax, NS, Canada, 2019.
5. Hoppe, A.; Mouriño, V.; Boccaccini, A.R. Therapeutic inorganic ions in bioactive glasses to enhance bone formation and beyond. *Biomater. Sci.* **2013**, *1*, 254–256. [CrossRef]
6. Mauro, J.C.; Philip, C.S.; Vaughn, D.J.; Pambianchi, M.S. Glass science in the United States: Current status and future directions. *Int. J. Appl. Glass Sci.* **2014**, *5*, 2–15. [CrossRef]
7. Hench, L.L.; Clark, D.E. Physical chemistry of glass surfaces. *J. Non-Cryst. Solids* **1978**, *28*, 83–105. [CrossRef]
8. Frugier, P.; Gin, S.; Minet, Y.; Chave, T.; Bonin, B.; Godon, N.; Lartigue, J.-E.; Jollivet, P.; Ayral, A.; De Windt, L.; et al. SON68 nuclear glass dissolution kinetics: Current state of knowledge and basis of the new GRAAL model. *J. Nucl. Mater.* **2008**, *380*, 8–21. [CrossRef]
9. Hellmann, R.; Cotte, S.; Cadel, E.; Malladi, S.; Karlsson, L.S.; Lozano-Perez, S.; Cabié, M.; Seyeux, A. Nanometre-scale evidence for interfacial dissolution-reprecipitation control of silicate glass corrosion. *Nat. Mater.* **2015**, *14*, 307–311. [CrossRef]
10. Huang, W.; Day, D.E.; Kittiratanapiboon, K.; Rahaman, M.N. Kinetics and mechanisms of the conversion of silicate (45S5), borate, and borosilicate glasses to hydroxyapatite in dilute phosphate solutions. *J. Mater. Sci. Mater. Med.* **2006**, *17*, 583–596. [CrossRef] [PubMed]
11. Wright, A.C. My Borate Life: An Enigmatic Journey. *Int. J. Appl. Glass Sci.* **2015**, *6*, 45–63. [CrossRef]
12. Yao, A.; Wang, D.; Huang, W.; Fu, Q.; Rahaman, M.N.; Day, D.E. In vitro bioactive characteristics of borate-based glasses with controllable degradation behavior. *J. Am. Ceram. Soc.* **2007**, *90*, 303–306. [CrossRef]

13. Ouis, M.; Abdelghany, A.; Elbatal, H. Corrosion mechanism and bioactivity of borate glasses analogue to Hench's bioglass. *Process. Appl. Ceram.* **2012**, *6*, 141–149. [CrossRef]
14. Bengisu, M. Borate glasses for scientific and industrial applications: A review. *J. Mater. Sci.* **2016**, *51*, 2199–2242. [CrossRef]
15. Biscoe, J.; Warren, B.E. X-Ray Diffraction Study of Soda-Boric Oxide Glass. *J. Am. Ceram. Soc.* **1938**, *21*, 287–293. [CrossRef]
16. Varshneya, A.K. *Fundamentals of Inorganic Glassmaking*; Elsevier: Amsterdam, The Netherlands, 1994. [CrossRef]
17. Maine, E.; Seegopaul, P. Accelerating advanced-materials commercialization. *Nat. Mater.* **2016**, *15*, 487–491. [CrossRef]
18. Arango-ospina, M.; Ilyas, K.; Boccaccini, A.R. Clinical Products and Further Devices of Phosphates and Borates. In *Phosphate and Borate Bioactive Glasses*; Royal Society of Chemistry: London, UK, 2022.
19. de Pablo, J.J.; Jackson, N.E.; Webb, M.A.; Chen, L.-Q.; Moore, J.E.; Morgan, D.; Jacobs, R.; Pollock, T.; Schlom, D.G.; Toberer, E.S.; et al. New frontiers for the materials genome initiative. *npj Comput. Mater.* **2019**, *5*, 41. [CrossRef]
20. Arróyave, R.; McDowell, D.L. Systems Approaches to Materials Design: Past, Present, and Future. *Annu. Rev. Mater. Res.* **2019**, *49*, 103–126. [CrossRef]
21. Cao, B.; Adutwum, L.A.; Oliynyk, A.O.; Luber, E.J.; Olsen, B.C.; Mar, A.; Buriak, J.M. How to Optimize Materials and Devices via Design of Experiments and Machine Learning: Demonstration Using Organic Photovoltaics. *ACS Nano* **2018**, *12*, 7434–7444. [CrossRef]
22. Axelrod, S.; Schwalbe-Koda, D.; Mohapatra, S.; Damewood, J.; Greenman, K.P.; Gómez-Bombarelli, R. Learning Matter: Materials Design with Machine Learning and Atomistic Simulations. *Accounts Mater. Res.* **2022**, *3*, 343–357. [CrossRef]
23. Jain, A.; Ong, S.P.; Hautier, G.; Chen, W.; Richards, W.D.; Dacek, S.; Cholia, S.; Gunter, D.; Skinner, D.; Ceder, G.; et al. Commentary: The materials project: A materials genome approach to accelerating materials innovation. *APL Mater.* **2013**, *1*, 011002. [CrossRef]
24. Piepel, G.F.; Cornell, J.A.; Piepel, G.F.; Cornell, J.A. Mixture Experiment Approaches: Examples, Discussion, and Recommendations Mixture Experiment Approaches: Examples, Discussion, and Recommendations. *J. Qual. Technol.* **1994**, *26*, 177–196. [CrossRef]
25. Bowden, G.D.; Pichler, B.J.; Maurer, A. A Design of Experiments (DoE) Approach Accelerates the Optimization of Copper-Mediated F-Fluorination Reactions of Arylstannanes. *Sci. Rep.* **2019**, *9*, 1–10. [CrossRef] [PubMed]
26. Tye, H. Application of statistical 'design of experiments' methods in drug discovery. *Drug Discov. Today* **2004**, *9*, 485–491. [CrossRef] [PubMed]
27. Murray, P.M.; Tyler, S.N.G.; Moseley, J.D. Beyond the Numbers: Charting Chemical Reaction Space. *Org. Process Res. Dev.* **2013**, *17*, 40–46. [CrossRef]
28. Kehoe, S.; Ardhaoui, M.; Stokes, J. Design of Experiments Study of Hydroxyapatite Synthesis for Orthopaedic Application Using Fractional Factorial Design. *J. Mater. Eng. Perform.* **2011**, *20*, 1423–1437. [CrossRef]
29. Anderson, M.J.; Whitcomb, P.J. *RSM Simplified: Optimizing Processes Using Response Surface Methods for Design of Experiments*, 2nd ed.; Productivity Press: New York, NY, USA, 2016.
30. Manchester, R.A.; Todorova, T.Z.; Werner-zwanziger, U.; Boyd, D. Mixture designs to investigate the role of alkali and alkaline earth cations on composition-structure-property relationships in ternary borate glass networks. *J. Non-Cryst. Solids* **2021**, *569*, 120982. [CrossRef]
31. ASTM International. ASTM E11: Standard Specification for Woven Wire Test Sieve Cloth and Test Sieves. Available online: https://www.astm.org/e0011-09e01.html (accessed on 18 April 2024).
32. Connell, K.O.; Hanson, M.; Shea, H.O.; Boyd, D. Linear release of strontium ions from high borate glasses via lanthanide/alkali substitutions. *J. Non-Cryst. Solids* **2015**, *430*, 1–8. [CrossRef]
33. ASTM International. *Standard Test Method for Assignment of the Glass Transition Temperatures by Differential Scanning Calorimetry*; ASTM International: West Conshohocken, PA, USA, 2008; Volume 8, pp. 1–4.
34. Topper, B.; Moncke, D.; Youngman, R.E.; Valvi, C.; Kamitsos, E.; Varsamis, C.P.E. Zinc borate glasses: Properties, structure and modelling of the composition-dependence of borate speciation. *Phys. Chem. Chem. Phys.* **2023**, *25*, 5967–5988. [CrossRef]
35. Alajerami, Y.S.M.; Hashim, S.; Hassan, W.M.S.W.; Ramli, A.T. The effect of titanium oxide on the optical properties of lithium potassium borate glass. *J. Mol. Struct.* **2012**, *1026*, 159–167. [CrossRef]
36. Kwinda, T.I.; Enke, D.; Koppka, S. Effect of TiO_2 on microphase development during phase separation and crystallization in Na2O–B2O3–SiO2 glass system. *J. Am. Ceram. Soc.* **2022**, *105*, 3261–3278. [CrossRef]
37. Anghel, E.M.; Florian, P.; Bessada, C. Structural Description of the $Na_2B_4O_7$-Na_3AlF_6-TiO_2 System. 1. IR and Raman Study of the Solidified Melts. *J. Phys. Chem.* **2007**, *111*, 962–967. [CrossRef] [PubMed]
38. Marzouk, M.A.; ElBatal, F.H.; ElBatal, H.A. Effect of TiO_2 on the optical, structural and crystallization behavior of barium borate glasses. *Opt. Mater.* **2016**, *57*, 14–22. [CrossRef]
39. Naresh, P.; Raju, G.N.; Kumar, V.R.; Piasecki, M.; Kiytyk, I.V.; Veeraiah, N. Optical and dielectric features of zinc oxy fluoro borate glass ceramics with TiO_2 as crystallizing agent. *Ceram. Int.* **2014**, *40*, 2249–2260. [CrossRef]
40. Kajinami, A.; Harada, Y.; Inoue, S.; Deki, S.; Umesaki, N. The Structural Analysis of Zinc Borate Glass by Laboratory EXAFS and X-Ray Diffraction Measurements. *Jpn. J. Appl. Phys.* **1999**, *38*, 132. [CrossRef]
41. Ehrt, D. Zinc and manganese borate glasses—Phase separation, crystallisation, photoluminescence and structure. *Phys. Chem. Glas. Eur. J. Glass Sci. Technol. Part B* **2013**, *54*, 65–75.
42. Yao, Z.; Möncke, D.; Kamitsos, E.; Houizot, P.; Célarié, F.; Rouxel, T.; Wondraczek, L. Structure and mechanical properties of copper-lead and copper-zinc borate glasses. *J. Non-Cryst. Solids* **2016**, *435*, 55–68. [CrossRef]

43. Colak, S.C.; Akyuz, I.; Atay, F. On the dual role of ZnO in zinc-borate glasses. *J. Non-Cryst. Solids* **2016**, *432*, 406–412. [CrossRef]
44. El-Falaky, G.E.; Guirguis, O.W. Effect of zinc on the physical properties of borate glasses. *J. Non-Cryst. Solids* **2012**, *358*, 1746–1752. [CrossRef]
45. Heuser, L.; Nofz, M. Journal of Non-Crystalline Solids: X Alkali and alkaline earth zinc and lead borate glasses: Structure and properties. *J. Non-Cryst. Solids X* **2022**, *15*, 100109. [CrossRef]
46. Shafaghi, R.; Rodriguez, O.; Phull, S.; Schemitsch, E.H.; Zalzal, P.; Waldman, S.D.; Papini, M.; Towler, M.R. Effect of TiO_2 doping on degradation rate, microstructure and strength of borate bioactive glass scaffolds. *Mater. Sci. Eng. C* **2020**, *107*, 110351. [CrossRef]
47. ElBatal, F.H.; Marzouk, M.A.; ElBatal, H.A. Crystallization and spectroscopic characterizations of binary SrO-B_2O_3 glasses doped with LiF, NaF, CaF_2, or TiO_2. *J. Aust. Ceram. Soc.* **2019**, *55*, 1039–1049. [CrossRef]
48. Hordieiev, Y.S.; Zaichuk, A.V. Effect of the addition of Al_2O_3, ZnO and TiO_2 on the crystallization behavior, thermal and some physical properties of lead borate glasses. *MRS Adv.* **2023**, *8*, 201–206. [CrossRef]
49. Song, S.; Wen, Z.; Liu, Y.; Zhang, Q.; Wu, X.; Zhang, J.; Han, J. Influence of dopants on the crystallization of borosilicate glass. *Ceram. Int.* **2009**, *35*, 3037–3042. [CrossRef]
50. Kržmanc, M.M.; Došler, U.; Suvorov, D. Effect of a TiO_2 nucleating agent on the nucleation and crystallization behavior of MgO-B_2O_3-SiO_2 glass. *J. Am. Ceram. Soc.* **2012**, *95*, 1920–1926. [CrossRef]
51. Rodriguez, O.; Curran, D.J.; Papini, M.; Placek, L.M.; Wren, A.W.; Schemitsch, E.H.; Zalzal, P.; Towler, M.R. Characterization of silica-based and borate-based, titanium-containing bioactive glasses for coating metallic implants. *J. Non-Cryst. Solids* **2016**, *433*, 95–102. [CrossRef]
52. Florian, P.; Anghel, E.M.; Bessada, C. Structural Description of the $Na_2B_4O_7-Na_3AlF_6-TiO_2$ System. 2. A Multinuclear NMR Approach of Melts and Solids. *J. Phys. Chem. B* **2007**, *111*, 968–978. [CrossRef] [PubMed]
53. Lakshmi, N.R.; Cole, S. Influence of TiO_2 ions on spectroscopic properties of oxyfluoride glasses. *Mater. Today Proc.* **2019**, *18*, 192–206. [CrossRef]
54. Lusvardi, G.; Malavasi, G.; Cortada, M.; Menabue, L.; Menziani, M.C.; Pedone, A.; Segre, U. Elucidation of the structural role of fluorine in potentially bioactive glasses by experimental and computational investigation. *J. Phys. Chem. B* **2008**, *112*, 12730–12739. [CrossRef] [PubMed]
55. Yin, T.; Lepry, W.C.; Hudon, P.; Ouzilleau, P.; Waters, K.E.; Nazhat, S.N. Mechanism of dissolution reactivity and reactions of various calcium borate glasses and glass-ceramics. *J. Non-Cryst. Solids* **2023**, *614*, 122406. [CrossRef]
56. Wu, J.; Stebbins, J.F.; Rouxel, T. Cation Field Strength Effects on Boron Coordination in Binary Borate Glasses. *J. Am. Ceram. Soc.* **2014**, *97*, 2794–2801. [CrossRef]
57. Hench, L.L.; Splinter, R.J.; Allen, W.C.; Greenlee, T.K. Bonding mechanisms at the interface of ceramic prosthetic materials. *J. Biomed. Mater. Res.* **1971**, *5*, 117–141. [CrossRef]
58. Yu, Y.; Stevensson, B.; Ede, M. Structural Role of Sodium in Borosilicate, Phosphosilicate, and Borophosphosilicate Glasses Unveiled by Solid-State NMR and MD Simulations. *J. Phys. Chem. C* **2019**, *123*, 25816–25832. [CrossRef]
59. Berkemeier, F.; Voss, S. Molar volume, glass-transition temperature, and ionic conductivity of Na- and Rb-borate glasses in comparison with mixed Na–Rb borate glasses. *J. Non-Cryst. Solids* **2005**, *351*, 3816–3825. [CrossRef]
60. Ahlawat, J.; Pawaria, S.; Bala, M.; Dahiya, S.; Ohlan, A.; Punia, R.; Maan, A.S. Materials Today: Proceedings Study of thermal and physical properties of sodium modified zinc borate glasses. *Mater. Today Proc.* **2023**, *79*, 118–121. [CrossRef]

Disclaimer/Publisher's Note: The statements, opinions and data contained in all publications are solely those of the individual author(s) and contributor(s) and not of MDPI and/or the editor(s). MDPI and/or the editor(s) disclaim responsibility for any injury to people or property resulting from any ideas, methods, instructions or products referred to in the content.

MDPI AG
Grosspeteranlage 5
4052 Basel
Switzerland
Tel.: +41 61 683 77 34

Materials Editorial Office
E-mail: materials@mdpi.com
www.mdpi.com/journal/materials

Disclaimer/Publisher's Note: The title and front matter of this reprint are at the discretion of the Guest Editor. The publisher is not responsible for their content or any associated concerns. The statements, opinions and data contained in all individual articles are solely those of the individual Editor and contributors and not of MDPI. MDPI disclaims responsibility for any injury to people or property resulting from any ideas, methods, instructions or products referred to in the content.

www.ingramcontent.com/pod-product-compliance
Lightning Source LLC
Chambersburg PA
CBHW062049220926
51850CB00010B/2707
9783725861354